Advances in 21st Century Human Settlements

Indexed by SCOPUS

This Series focuses on the entire spectrum of human settlements – from rural to urban, in different regions of the world, with questions such as: What factors cause and guide the process of change in human settlements from rural to urban in character, from hamlets and villages to towns, cities and megacities? Is this process different across time and space, how and why? Is there a future for rural life? Is it possible or not to have industrial development in rural settlements, and how? Why does 'urban shrinkage' occur? Are the rural areas urbanizing or is that urban areas are undergoing 'ruralisation' (in form of underserviced slums)? What are the challenges faced by 'mega urban regions', and how they can be/are being addressed? What drives economic dynamism in human settlements? Is the urban-based economic growth paradigm the only answer to the quest for sustainable development, or is there an urgent need to balance between economic growth on one hand and ecosystem restoration and conservation on the other – for the future sustainability of human habitats? How and what new technology is helping to achieve sustainable development in human settlements? What sort of changes in the current planning, management and governance of human settlements are needed to face the changing environment including the climate and increasing disaster risks? What is the uniqueness of the new 'socio-cultural spaces' that emerge in human settlements, and how they change over time? As rural settlements become urban, are the new 'urban spaces' resulting in the loss of rural life and 'socio-cultural spaces'? What is leading the preservation of rural 'socio-cultural spaces' within the urbanizing world, and how? What is the emerging nature of the rural-urban interface, and what factors influence it? What are the emerging perspectives that help understand the human-environment-culture complex through the study of human settlements and the related ecosystems, and how do they transform our understanding of cultural landscapes and 'waterscapes' in the 21st Century? What else is and/or likely to be new vis-à-vis human settlements – now and in the future? The Series, therefore, welcomes contributions with fresh cognitive perspectives to understand the new and emerging realities of the 21st Century human settlements. Such perspectives will include a multidisciplinary analysis, constituting of the demographic, spatio-economic, environmental, technological, and planning, management and governance lenses.

If you are interested in submitting a proposal for this series, please contact the Series Editor, or the Publishing Editor:

Bharat Dahiya (bharatdahiya@gmail.com) or
Loyola D'Silva (loyola.dsilva@springer.com)

Seth Asare Okyere · Matthew Abunyewah ·
Michael Odei Erdiaw-Kwasie ·
Festival Godwin Boateng
Editors

Urban Slums and Circular Economy Synergies in the Global South

Theoretical and Policy Imperatives for Sustainable Communities

Springer

Editors
Seth Asare Okyere
School of Landscape Architecture
and Planning
University of Arizona
Tucson, AZ, USA

Michael Odei Erdiaw-Kwasie
Asia Pacific Center for Business
and Sustainable
Charles Darwin University
Darwin, NT, Australia

Matthew Abunyewah
College of Health Sciences
Charles Darwin University
Darwin, NT, Australia

Festival Godwin Boateng
School of Geography and the Environment
Oxford University
Oxford, UK

ISSN 2198-2546 ISSN 2198-2554 (electronic)
Advances in 21st Century Human Settlements
ISBN 978-981-99-9024-5 ISBN 978-981-99-9025-2 (eBook)
https://doi.org/10.1007/978-981-99-9025-2

This Springer imprint is published by the registered company Springer Nature Singapore Pte Ltd.
The registered company address is: 152 Beach Road, #21-01/04 Gateway East, Singapore 189721, Singapore

Paper in this product is recyclable.

Foreword by Prof. Thayaparan Gajendran

The editors and authors of this book have done fabulous work in addressing a complex and imminent issue in a profound way—providing a powerful perspective into sustainability and social justice in the context of slums in the Global South—through a circular economy perspective. This book demonstrates the contributors' passion for deliberating wide-ranging issues impacting slum dwellers and the nexus between slums and the circular economy.

The Global South countries have a greater number of slums with lower levels of development. Most of these countries are impacted by historical colonialism and exploitation and often face issues such as poverty, lack of access to education and healthcare, and low levels of human development. These slums are densely populated with poor living conditions stemming from substandard housing and inadequate basic infrastructure. Over the decades, the population living in slums in these countries has not significantly improved their conditions—however, it has had increased exposure to risks.

Historically the focus on slums was on policy issues concerning unplanned urbanization, affordable housing, and access to essential services. The primary discourse was on rapidly growing cities and how governments, non-governmental organizations, and community organizations can assist in improving the living conditions in slums—mainly through better housing, infrastructure, and social services. The focus was on comprehensive urban planning, affordable housing solutions, and strategies to promote economic and social development.

Historically, the slums are seen as impediments to sustainability and sustainable development. The basic developmental challenges faced by people living in slums have undermined their ability to enhance their sustainability and circularity practices that can improve their engagement with the mainstream economy while contributing toward the SDGs. Moreover, it can be said that development challenges have increased the environmental and disaster risks in slums in the way of impacts from hazardous waste and exposure, increasing the intensity of natural hazards.

This book has an intriguing collection of chapters that flips the discourse about slums from a policy challenge to policy opportunities for development through a circular economy—emerging the productive synergies that can be built in urban slums and circular economy. That is engaging slums in a circular economy and making them a system that minimizes waste, and emission, closing and narrowing material and energy loops. This book argues an urgent need to document and theorize the slum–circular economy nexus more systematically and holistically and situate slums—livelihoods, housing, and spaces (natural and built space) for illuminating the synergies with circular economy principles. By doing this, identify the key developmental and sustainability challenges and use the circular economy to address them.

This book illuminates the nexus between the slums and circular economy relating to architecture and housing, waste practices (e-waste, construction, and demonization waste), circular ways through participatory urban planning, and energy generation through agricultural residues. The book covers perspectives assembled from empirical and review cases from Sub-Saharan Africa, South Asia, and Latin America.

This book provides an excellent addition to knowledge on circular economy and slums in Global South and a valuable contribution to theory and practice—the alternative discourse to addressing challenges in slums through the circular economy is a must-read for scholars and practitioners.

Prof. Thayaparan Gajendran
Associate Professor, School
of Architecture and Built Environment
University of Newcastle
Callaghan, Australia

Foreword by Prof. Michihiro Kita

We are experiencing a period of dramatic change for both the world and local communities. While significant scientific and technological progress is being made, we must cope with various complex socio-environmental issues and potentially existential threats such as climate change and global pandemics. Globally, there are major shifts in the social, economic, and spatial systems that humanity relies upon. From an urban perspective, humankind is still grappling with the specific form or idea for cities that can respond to these changes. Progress toward sustainable and inclusive urban development calls for historical lessons and the need to decenter global learning and practices from the global north to account for southern experiences and practices.

From the *context* of cities and communities, there are two major turning points in the modernization process of urban planning. The first was from the end of the nineteenth century to the beginning of the twentieth century, after the implementation of the urban renewal of Paris by Georges-Eugène Haussmann, followed by a trend of major remodeling in other European cities. Various urban planning ideas were presented to resist this trend. For example, Camillo Sitte explained the principle of the beauty of urban organization, which was self-organized over hundreds of years, and criticized the inorganic nature of the geometric urban forms that were oriented at the time (Sitte 1889). Patrick Geddes also presented the city as a living organism and the importance of understanding with citizens the process of evolution as a vital form from the past to the present day and reflecting it in planning to promote the development of the city (Geddes 1915). The intellectual accumulation of the ideas that tried to protect the "context" of the city might be considered the first wave of contextualism (Architectural Institute of Japan 2022).

However, after World War II, against the background of economic growth, large-capital urban development that renewed existing cities spread all over Europe, the United States, and other regions that are considered developed countries. To strongly object to this, Robert Venturi and Colin Lowe believed that it is important for humanity to constantly evolve its "context" in a creative way, which can be achieved by developing rather than destroying traditions (Venturi 1966, Lowe and Koetter, 1983, Akimoto 2002). Scholars like Christopher Alexander (1977, 1979) and

Jane Jacobs (1961) presented theories highlighting the value of context and people attributes/practices essential for building better cities. This intellectual convergence that emerged from the 1960s through the early 1980s can also be viewed as the second wave of contextualism. Since then, theories and ideas related to urban planning have developed by going back and forth between destruction/renewal and inheritance/development of the *context*. Today, in the background of global capitalism and neoliberalism, large-scale constructions that destroy the *context occur in* cities considered developing countries. The challenge imposed upon scholars and practitioners is how to counter this destruction and create a third wave of contextualism.

What themes should be carried over from the first and second waves to the third? Here, this book is instructive in its contribution: weaving together everyday slum life with the economy. Drawing on this, without romanticizing, the power of self-organized living is a theme that should remain central. One theme that needs further progress is the "economy." I speculate on the creation of small circular economic activities in urban spaces. One thread of ideas that should be constructed in the third wave would be the organization of diverse units of economic activity, large, medium, and small, as a continuous organism of upper and lower levels, as Alexander suggested for spatial planning. This would work as a principle to form a symbiotic society where diverse people find their roles and enhance their economic well-being through circular mechanisms.

The first wave of contextualism occurred in Europe, and the starting point of the second wave was the United States. I believe the third wave will emerge from the Global South. It is hoped that the findings in this book will be the nucleus for a creative evolution of planning, sustainability, and circular economy ideas in the world by "translating" the organic circular economy practices/mechanisms from the Global South to other world cities which have lost the emphasis on what the urban marginalized people are already doing in the so-called circular city pathways.

Prof. Michihiro Kita
Director, Social Solutions Initiative,
Office of the Vice President
Osaka University
Suita, Japan

Preface

In the Preface to the book, a few lines explaining the origin of this book wouldn't be out of place. Nature Sustainability Journal featured our article on the circular economy-slum model that we have been working on for a few years. Following up on this ground-breaking publication, we were interested in expanding our knowledge by proposing a book to Springer Nature on the subject. After considerable revision and additions, the proposed edited book has come to fruition with quality contributions from authors from different developing regions.

Over the years, we have come to appreciate the difficulty our peers face when trying to learn about the opportunities associated with slums in our growing cities. The available books mostly focus on socio-economic and environmental problems associated with slums than anything else. The space of slum research has evolved and transformed from a relatively inward-looking view of slum pathogenicity to an outward-looking view that seeks opportunities. With the rising optimism about the social, economic, and environmental prospects that the circular economy holds for many emerging and developing economies, it is time to enrich the macro-level analysis of the circular economy by incorporating the slum aspect of the debate. We believe that for successful inclusive city planning to be realized, policymakers, practitioners, and academics do not need a narrow perspective of the slum story. Instead, they need a working understanding of what questions to ask and when to ask, what approaches to employ to get the answers, and how to interpret the results correctly.

The book is structured to make circular economy-slum research interactive and easy to understand. We have used illustrations to communicate with our readers and present cutting-edge research across different developing country contexts. The book combines essential concepts with classic and current case studies illustrating a macro-level analysis of the synergies from different countries. The chapters are thorough, balanced in coverage, and bite-sized in length so that readers do not lose grip of their attention span. For a learner, the book's emphasis is to lead you beyond your existing understanding of circular economy-slum synergies, guide you to apply foundational concepts to contemporary issues and headlines, and ask questions along the way. For academics, the book is intended as a guide and hopes to provide you with enough theoretical and empirical foundations to navigate the maze of circular economy-slum research. For policymakers and practitioners, the book offers significant developments in recent studies into practice.

Tucson, USA — Seth Asare Okyere
Darwin, Australia — Matthew Abunyewah
Darwin, Australia — Michael Odei Erdiaw-Kwasie
Oxford, UK — Festival Godwin Boateng

Acknowledgments

This book elevates the agency of slum dwellers and residents of informal settlements and their everyday activities that intersect with the core principles of the circular economy—a topic that has gained traction in the sustainability transition discourse but is rarely considered from the perspective of Global South urbanism. Therefore, our contributors deserve mention for their contribution to this theme and for giving empirical and theoretical visibility to this topic that has remained in the blind spots of scholarship, policy, and practice for so long. We say thank you for helping us advance what was once an idea—just a thought. We are thankful to the editorial assistants at Springer Nature for their support in bringing the text home and sharpening it. Stephen Leonard Mensah at the University of Memphis deserves special mention for his invaluable assistance to Seth in preparing this book for production.

The editorial team expresses our gratitude to all the reviewers of the chapters for their contribution to the quality of this book through their comments and suggestions of earlier drafts. To Prof. Gajen and Prof. Kita, we say thank you for kindly and excitedly agreeing to pen the foreword. Finally, we thank the editors at Nature Sustainability, who published our perspective paper "*Advancing slum–CE model for sustainability transition in Global South cities*"—an important paper that set the foundation for this book.

We dedicate this book to all the slum and marginalized informal dwellers of Global South cities, who, through their improvisations and everyday struggles, strife to make something out of nothing; contest, defy, persist, and endure unequal, unjust, and uneven practices in the so-called building of the sustainable city.

Contents

Introduction: The Embeddedness of Circularity in Everyday Slum Living in Global South Cities 1
Seth Asare Okyere, Festival Godwin Boateng, Matthew Abunyewah, and Michael Odei Erdiaw-Kwasie

Linking Informal Settler Practices and Circular Principles in Global South: Lessons from Asian and Latin American Cities 11
Isaac Nevis Fianoo, Stephen Leonard Mensah, Seth Asare Okyere, Matthew Boadie-Ampong, and Joshua Jirjiri

Modern Vernacular Architecture and Circular Economy in Informal Settlements 27
Johan Mottelson

Exploring Circular Economy Awareness, Perceptions and Practices in Selected Urban Slums in Kigali City of Rwanda 47
Vanessa Robertson, Eucabeth Majiwa, and Nathan Kanuma Taremwa

Recycling from Construction and Demolition (C&D) Waste: Exploring the Scope of Circular Management in Constructing Slum Dwellings of Dhaka 71
Md. Nawrose Fatemi and Tahmina Rahman

Circularities in Housing Transformation Practices: A Synergetic Review 95
Florence Avogo Abugtane, Seth Opoku Mensah, and Gervase Kuuwaabong

Circular Economy in Africa's Informal Cities: A Review of Residents' Value Retention Practices and Their Implications for Participatory Urban Planning 113
Stephen Leonard Mensah, Louis Kusi Frimpong, Seth Asare Okyere, Shine Francis Gbedemah, and Matthew Abunyewah

Circularizing Livelihoods: Transforming Agricultural Residues to Electricity in Low-Income Peri-Urban Areas of Uganda 131
Nataliya Apanovich

Metropolitan Cartography: A Novel Approach for Assessing How New Morpho-Types Solutions Impact the Circular City Agenda. The Ouagadougou Case Study 149
Antonella Contin and Alessandra Pandolfi

Closing the Policy-Implementation Gaps in E-Waste Management: Implications for Circular Economy and Sustainability in Urban Ghana 173
Louis Kusi Frimpong, Stephen Leonard Mensah, and Seth Asare Okyere

Towards just Circular Transitions in the Slums of Global South Cities 193
Matthew Abunyewah, Seth Asare Okyere, Michael Odei Erdiaw-Kwasie, and Festival Godwin Boateng

Index 205

Editors and Contributors

About the Editors

Dr. Seth Asare Okyere is an interdisciplinary development planner working at the intersection of international development, sustainability, resilience, and social equity. He has extensive experience across academia, policy agencies, and non-governmental organizations conducting transformative research, leading international expert teams, coordinating international joint projects, and engaging local communities to build capacity for sustainable development. His experiences span cities and communities in Africa, Europe, Asia, and North America. Seth is in pursuit of the sustainable development of vulnerable communities within mission-focused approaches to co-producing equitable, resilient, and place-based solutions. Seth holds a Ph.D. in Urban Development Planning (Osaka University, Japan) and an M.Sc. in Urban Planning and Policy Design (Polytechnic University of Milan, Italy). He currently serves as Visiting Assistant Professor (University of Arizona, USA) and Adjunct Associate Professor (Osaka University, Japan).

Dr. Matthew Abunyewah is a research-focused lecturer at the Australasian Centre for Resilience Implementation for Sustainable Communities within the Faculty of Health and Adjunct Senior Lecturer at the University of Newcastle, Australia. He received his Ph.D. in Disaster Management and Resilience from the University of Newcastle, Australia. Matthew leverages participatory and interdisciplinary research approaches to inform government policies on disaster management and resilience, circular economy and sustainability, community development, industry and business development, and climate change and variability. Before joining Charles Darwin University, Dr. Abunyewah worked in the private and public sectors, contributing to policy formulation and implementation in Ghana and Australia. For example, Matthew contributed to developing the Northern Territory Circular Economy Strategy 2022–2027.

Dr. Michael Odei Erdiaw-Kwasie is an award-winning author and Lecturer in the Business and Accounting Discipline at Charles Darwin University. Dr. Erdiaw-Kwasie holds a Ph.D. in Sustainability Management from the University of Southern Queensland. His research interests focus on sustainability transitions, circular economy, climate risks and resilience, and global development. He has published over 40 scholarly outputs—books, book chapters, journal articles, and discussion papers in various fields. Dr. Erdiaw-Kwasie is Editorial Board Advisory Member for the Business Strategy and the Environment Journal. Dr. Erdiaw-Kwasie has worked across policy agencies and international organizations such as Transparency International, UNDP, and Intergovernmental Forum (IGF) in various capacities.

Dr. Festival Godwin Boateng is a political economist. His research interests are diverse and focus on engaged scholarship at the intersection of transport, housing, land use, pollution, climate, urban governance, and policy. Festival formerly worked as Postdoctoral Research Scholar at the Climate School of Columbia University, New York, and as Adjunct Professor at the Department of Urban Policy and Planning of Hunter College, New York. He currently serves on the boards of the Consortium for Sustainable Urbanization and Nature: Humanities and Social Sciences Communications and will be joining the Transport Studies Unit of the University of Oxford, UK, in the winter of 2023 as Senior Researcher in Mobility Governance. Festival received his Ph.D., M.Sc., and B.A. (Hons.) degrees from the Royal Melbourne Institute of Technology (RMIT), Melbourne-Australia, Roskilde University, Copenhagen-Denmark, and Kwame Nkrumah University of Science and Technology, Kumasi-Ghana, respectively.

Contributors

Florence Avogo Abugtane Department of Planning, SD Dombo University of Business and Integrated Development Studies, Wa, Ghana

Matthew Abunyewah The Australian Center for Resilience Implementation for Sustainable Communities, Charles Darwin University, Darwin, Australia

Nataliya Apanovich School of Landscape Architecture and Planning, University of Arizona, Tucson, AZ, USA

Seth Asare Okyere School of Landscape Architecture and Planning, The University of Arizona, Tucson, USA

Matthew Boadie-Ampong Institute for the Advanced Study of Sustainability, United Nations University, Tokyo, Japan

Festival Godwin Boateng The Earth Institute, Columbia University, New York, USA;
School of Geography and the Environment, University of Oxford, Oxford, UK

Antonella Contin Politecnico di Milano, Dipartimento di Architettura e Studi Urbani, Milan, Italy

Michael Odei Erdiaw-Kwasie Business and Accounting Discipline, Faculty of Arts & Society, Charles Darwin University, Casuarina, NT, Australia

Md. Nawrose Fatemi Department of Architecture, University of Asia Pacific (UAP), Dhaka, Bangladesh

Isaac Nevis Fianoo Institute for the Advanced Study of Sustainability, United Nations University, Tokyo, Japan

Shine Francis Gbedemah Department of Geography and Earth Sciences, University of Environment and Sustainable Development, Somanya, Ghana

Louis Kusi Frimpong Department of Geography and Earth Science, University of Environment and Sustainable Development, Somanya, Ghana

Joshua Jirjiri Department of Development Management, SD Dombo University of Business and Integrated Development Studies, Wa, Ghana

Gervase Kuuwaabong Department of Geography, SD Dombo University of Business and Integrated Development Studies, Wa, Ghana

Stephen Leonard Mensah Department of City and Regional Planning, The University of Memphis, Memphis, TN, USA

Eucabeth Majiwa Jomo Kenyatta University of Agriculture and Technology (JKUAT), Nairobi, Kenya

Seth Opoku Mensah Institute for Sustainable Futures, University of Technology Sydney, Ultimo, Australia

Johan Mottelson Institute of Architecture, Urbanism and Landscape, Royal Danish Academy, Copenhagen, Denmark

Alessandra Pandolfi Politecnico di Milano, Dipartimento di Architettura e Studi Urbani, Milan, Italy

Tahmina Rahman Department of Architecture, University of Asia Pacific (UAP), Dhaka, Bangladesh

Vanessa Robertson Neu-Ulm University of Applied Sciences (HNU), Neu-Ulm, Germany

Nathan Kanuma Taremwa University of Rwanda (UR), Kigali, Rwanda

List of Figures

Linking Informal Settler Practices and Circular Principles in Global South: Lessons from Asian and Latin American Cities

Fig. 1 Projections of Slum Population by 2025 (*Source* Adapted from the UN-Habitat 2018a, b) 12

Modern Vernacular Architecture and Circular Economy in Informal Settlements

Fig. 1 Location of the cities in which the case studies are situated 32
Fig. 2 An overview of the built environment of **a** Khâm Thiên (Hanoi, Vietnam), **b** Maxaquene (Maputo, Mozambique), and **c** Rocinha (Rio de Janeiro, Brazil) 33
Fig. 3 Typical street spaces in Khâm Thiên (left), Maxaquene (middle), and Rocinha (right) 34
Fig. 4 Location of Khâm Thiên in Hanoi, Vietnam 34
Fig. 5 Location of Maxaquene in Maputo, Mozambique 35
Fig. 6 Location of Rocinha in Rio de Janeiro, Brazil 36
Fig. 7 Construction sites in Khâm Thiên (left), Maxaquene (middle), and Rocinha (right) 37
Fig. 8 Adaptable design strategies in Rocinha 38
Fig. 9 **a** Makeshift structure partially made of reclaimed asbestos cement boards, **b** Collection of used plastics for recycling 39
Fig. 10 Downcycling of construction materials in Maxaquene (left) and Khâm Thiên (right) 40
Fig. 11 Upcycling of waste materials used for interior wall cladding in Khâm Thiên 42
Fig. 12 **a** Rental housing area in Khâm Thiên largely built of reclaimed materials, **b** Façade made of reclaimed materials, **c** Asbestos cement boards stored next to dwelling 43

Exploring Circular Economy Awareness, Perceptions and Practices in Selected Urban Slums in Kigali City of Rwanda

Fig. 1 Rwanda districts map (Reproduced from Government of Rwanda 2009) ... 53
Fig. 2 Key informants' level of education. *Source* Own research ... 54
Fig. 3 Key informants' self-indicated level of awareness of CE-related concepts. *Source* Own research ... 57
Fig. 4 KI's level of agreement to CE-attitude related statements. *Source* Own research ... 58
Fig. 5 KI's level of agreement to CE-knowledge related statements. *Source* Own research ... 59
Fig. 6 KIs' level of agreement to CE-behaviour related statements. *Source* Own research ... 62
Fig. 7 Motivations for CE-related activities according to the 15 KIIs. *Source* Own research ... 63

Recycling from Construction and Demolition (C&D) Waste: Exploring the Scope of Circular Management in Constructing Slum Dwellings of Dhaka

Fig. 1 Selected 5 sites in the map of slums of Dhaka City (CUS 2005); modified by Author ... 77

Circular Economy in Africa's Informal Cities: A Review of Residents' Value Retention Practices and Their Implications for Participatory Urban Planning

Fig. 1 Informal Settlements and urban circularity synergies (*Source* Author's elaboration) ... 123

Circularizing Livelihoods: Transforming Agricultural Residues to Electricity in Low-Income Peri-Urban Areas of Uganda

Fig. 1 Electricity potential by residue type ... 141
Fig. 2 Comparison of Uganda's current power capacity and electricity produced from technical potential and wasted agricultural residues ... 142

Metropolitan Cartography: A Novel Approach for Assessing How New Morpho-Types Solutions Impact the Circular City Agenda. The Ouagadougou Case Study

Fig. 1 Burkina Faso. Ouagadougou. Protocol Map. Ecosystem Analysis (MSLab) ... 151

Fig. 2 Burkina Faso. Ouagadougou, Metropolitan Architecture Project, Masterplan Özge Tuncay, MSLab, 2023, Master Thesis. Politecnico di Milano 152
Fig. 3 Burkina Faso. Ouagadougou, Metropolitan Architecture Project, Masterplan (MSLab, Özge Tuncay) 153
Fig. 4 Ouagadougou. Former airport area project. New Railway Station. New urban morphotype as linkage urban–rural pattern (MSLab & Marcello Longeri, Christodoulos Iosif) 154
Fig. 5 Ouagadougou. Koubri Green Axis Section (MSLab & Yalda Beigagha, Ilayda Comert, Kubra Ilayda Turan) 155
Fig. 6 Ouagadougou. Urban Section Landscape showing the connection from Centre to Koubri (MSLab & Yalda SSSBeigagha, Ilayda Comert, Kubra Ilayda Turan) 156
Fig. 7 Ouagadougou. Metropolitan Architecture Project Physical dimension able to sustain a circular economy project (MSLab & Ozge Tuncay) 159
Fig. 8 Ouagadougou. Innovative campus (MSLab & Prarthana Ulaganathan, Ekaterina Kovaleva) 163
Fig. 9 Ouagadougou. Metro-Matrix 169

Towards just Circular Transitions in the Slums of Global South Cities

Fig. 1 Conceptual scheme of urban slum-circular economy synergy. *Source* Authors' elaboration 197

List of Tables

Linking Informal Settler Practices and Circular Principles in Global South: Lessons from Asian and Latin American Cities

Table 1 Selected cases of informal settlements practices 15

Exploring Circular Economy Awareness, Perceptions and Practices in Selected Urban Slums in Kigali City of Rwanda

Table 1 Key informants' information 54
Table 2 Overview of focus group discussions 55
Table 3 Prevalence of circular practices in the three slum areas according to FDG results 61
Table 4 Challenges in implementing the CE according to the FGD findings 64

Recycling from Construction and Demolition (C&D) Waste: Exploring the Scope of Circular Management in Constructing Slum Dwellings of Dhaka

Table 1 Regulatory Documents for Solid Waste Management in Bangladesh 75
Table 2 Demographic profile of the respondents in the study areas 78
Table 3 Policy review on circular management of C&D waste in Dhaka 80
Table 4 Summary of the policy review 82
Table 5 Correlational Value of Attributes with Familiarity of Slum Dwellers 83
Table 6 Respondents' preferences and the age of their house 84
Table 7 Respondents' Preference and Construction Material Collection Procedure 85
Table 8 Respondents' preference and familiarity with recycled C&D waste 87

Table 9 Respondents' preference and quality of recycled C&D waste ... 88
Table 10 Respondents' preference and cost of recycled C&D waste 88
Table 11 Respondents' preference and durability of recycled C&D waste 89

Circularities in Housing Transformation Practices: A Synergetic Review

Table 1 Synergies between housing transformations and circular cities 107

Circularizing Livelihoods: Transforming Agricultural Residues to Electricity in Low-Income Peri-Urban Areas of Uganda

Table 1 Review and research article results using Boolean operators and wildcards 133
Table 2 Agricultural waste-to-energy technology applications and barriers to adoption in Uganda 134
Table 3 Techno-economic summary of Tiribogo village gasifier in central Uganda 137
Table 4 Agricultural residue uses by farming households in central Uganda (Masaka District) 138
Table 5 Selected crops and annual residue production in central Uganda in 2019 140

Metropolitan Cartography: A Novel Approach for Assessing How New Morpho-Types Solutions Impact the Circular City Agenda. The Ouagadougou Case Study

Table 1 Project phases 166
Table 2 Actions to be activated 167
Table 3 Criticalities to be solved 168
Table 4 Positive factors and outcomes of the project 168

Towards just Circular Transitions in the Slums of Global South Cities

Table 1 Synergies between circular economy and slum 195

Introduction: The Embeddedness of Circularity in Everyday Slum Living in Global South Cities

Seth Asare Okyere, Festival Godwin Boateng, Matthew Abunyewah, and Michael Odei Erdiaw-Kwasie

Abstract The reality of *informalization* and '*slummification*' of Global South cities has garnered much interest in academic and policy circles at international, national, and local levels, as reflected in the Sustainable Development Goals, the New Urban Agenda, and even broader issues around Sustainable Futures. Here, the Circular Economy (CE), which encompasses questions around material production, use, re/de-compositions, and their environmental impact, is gradually gaining a foothold in sustainability discourses in southern cities. Yet, the gradual insertion of CE in the Global South often perceives urban slum dweller activities and livelihoods as detrimental to the circular cities and sustainable futures agenda. This introductory chapter sets the context for rethinking this conventional pathological view of the urban slum–CE nexus. It briefly synthesizes urban transitions, circular economy, and urban slums. It then provides an overview of the collection of insightful essays in this volume that foreground the synergies between everyday activities in slums (livelihoods, housing, and space) and the core principles of the circular economy. In

---

S. A. Okyere (✉)
School of Landscape Planning and Architecture, University of Arizona, Tucson, USA
e-mail: sethaokyere@arizona.edu

F. G. Boateng
The Earth Institute, Columbia University, New York, USA

School of Geography and the Environment, University of Oxford, Oxford, UK

F. G. Boateng
e-mail: festival.boateng@ouce.ox.ac.uk

M. Abunyewah
Australian Center for Resilience Implementation and Sustainable Communities, Charles Darwin University, Casuarina, Australia
e-mail: matthew.abunyewah@cdu.edu.au

M. O. Erdiaw-Kwasie
Business and Accounting Discipline, Faculty of Arts & Society, Charles Darwin University, Casuarina, NT 0810, Australia
e-mail: michael.erdiaw-kwasie@cdu.edu.au

S. A. Okyere et al. (eds.), *Urban Slums and Circular Economy Synergies in the Global South*, Advances in 21st Century Human Settlements,
https://doi.org/10.1007/978-981-99-9025-2_1

doing so, it opens the door for systematic, pragmatic, and locally situated theoretical and policy directions to explore the role of urban slums in the circular cities agenda.

Keywords Urban slums · Circular economy · Global south · Urban sustainability

1 Setting the Context

The urban share of the global population has constantly increased over the past decades, from 29.6% in 1950 to 55% in 2013 and subsequently to 56.2% in 2020. This is expected to increase further to 60.4% and 66%, respectively, by 2030 and 2050, with about 96% of the growth projected to occur in the developing world (UN-Habitat, 2020). Currently, cities and towns are said to consume 60–80% of natural resources globally and produce 50% of global waste and 75% of greenhouse gas emissions (Williams, 2019), with sustainability transition practitioners and researchers warning that, without urgent corrective measures to the current rate and scale of natural resource depletion, global pollution, and socio-environmental harms, the planet's capacity to support socioeconomic livelihoods will be compromised irreversibly (Abunyewah et al., 2023; D'Odorico et al., 2018).

These concerns have stimulated calls for a production and consumption approach that has the capability of slowing and narrowing resource use through the prevention of waste, improvement of resource use, and substitution of the use of primary resources with recovered materials and energy. As a response and an alternative to the current unsustainable linear production and consumption paradigm, the circular economy (CE) concept has emerged, gaining momentum in urban studies (Christensen, 2021; Schröder et al., 2020).

Geissdoerfer et al. (2017) define CE as a regenerative system that minimizes waste, emission, and energy leakage by slowing, closing, and narrowing material and energy loops. Blomsma and Brennan (2017) indicate that CE is an umbrella concept underpinned by greater resource efficiency and effectiveness. Despite concerns about its definitional/conceptual/theoretical confusions, implementation challenges, and depoliticization of sustainable growth (see Corvellec et al., 2022), across diverse academic, policy, and activist circles, CE is perceived to have the potential to help resolve the highly complex challenges of simultaneously satisfying human needs while preserving the environment and tackling climate change challenges (Christensen, 2021; EMF, 2013).

Broadly, current approaches to CE focus on two main lines of inquiry: *state* and *business/firm* level interventions. The state or unions of state-level studies investigate what such actors do or are doing to stimulate societal shifts to CE at the global, regional, and national, as well as city levels while firm-based studies often focus on CE business models, strategies, or innovations (Bocken et al., 2016; Schwanholz and Leipold, 2020; Lüdeke-Freund et al., 2019). Bocken et al. (2016) distinguish two categories of business models or strategies that dominate CE research: *slowing resource* and *closing resource* loops. The first centers on the creation of products with longer life and the development of product reuse practices, while the second relates to creating value from what the linear model considers as waste (see Bocken et al., 2016; Stahel, 2016).

The heavy representation of businesses and states in CE research and policy is warranted; such high-level actors and others similarly situated are and will remain critical to the global development goal of shifting from the present take-make-consume-and-dispose approach to a more sustainable one. The downside, however, is that focusing so much on what states and businesses do or should do divests attention from exploring and building on the lessons from what *local everyday people* do or are already doing that also have the potential to support the sustainability transition. Some recent studies have begun to approach CE from a subaltern perspective, documenting the overlooked circularities in daily practices and local political economies. However, as with CE research generally, many of these studies largely take their cases from Global North countries (Kębłowski et al., 2020; Wuyts and Marin, 2022). Overall, much of the discussions about CE have overlooked Global South cities—and, in particular, the sprawling impoverished slums that run in and around the more built-up central business and industrial districts of the cities (Patel, 2023; Allam and Jones, 2018).

According to the UN-Habitat (2008), the concept of a 'slum' refers to urban areas with the following conditions: non-durable structures (e.g., shacks), insufficient living area (i.e., overcrowding), and deficient access to adequate water and sanitation facilities. In practice, however, slums can be hard to define precisely (Engstrom et al., 2019). Despite all the progress made in improving and preventing slum formation—represented by the decline in urban slum population in developing countries from 39 to 30% between 2000 and 2014—a lot more people continue to experience impeded access to basic necessities of life, including adequate serviced housing. For instance, the UN-Habitat 2015–2016 *Slum Almanac* puts the global slum population at around 1 billion people (UN-Habitat, 2016).

Slums play key roles, especially in the urban economies of developing countries. They provide affordable housing to the low-income urban working class who are normally excluded from the formal housing market (Obeng-Odoom, 2011; Klopp and Paller, 2019; Paller, 2014; 2015; 2019). For instance, about one in three urban residents (over 900 million people) in the Global South are said to reside in slums (Turok and Borel-Saladin, 2018). Further, slum dwellers' usually low economic and educational standings force them to take up low-paying jobs that may not be readily accepted by middle-and upper-class people (Okyere et al. 2021). More importantly, slum dwellers often find jobs in the informal sector and engage in activities—such as selling dog chains, scrap metals, and providing cheap labor—that indirectly subsidize the lifestyles of the middle and high-income class who access goods and services from slums extremely cheaply (Obeng-Odoom, 2013).

In effect, further contributing to the informal economy—an important sector of developing countries' economies—the activities of slum dwellers subsidize the lifestyles of the middle and high-income class and the operation costs of businesses, including formal ones. Thus, slums and the activities of slum dwellers contribute in great ways to oiling the urban economy. Yet, slums and slum dwellers are widely derided. They are perceived to carry significant societal, environmental, and other costs. Slums are frequently represented in the literature as settings of poverty, diseases, dysfunctionality, congestion, crime, and grime (Davis, 2013; Bob-Millar and Obeng-Odoom, 2011). The representation of the impact of slums and the activities of slum dwellers on the environment in the literature deserves more careful consideration.

Slum dwellers typically inhabit marginal locations, including riverbanks, steep slopes, or dumping grounds, thereby heightening cities' vulnerability to the damaging wrath of natural disasters (Abunyewah et al., 2018; Mahabir et al., 2016). The waste management challenges in slums are said to create room for diseases, pollution, and other environmental problems to thrive (see Obeng-Odoom, 2011; 2013). Overall, slums tend to be represented in environmental sustainability discourses as the exemplifying phenomenon that intertwines the socioeconomic and environmental problems of urbanization (see Abunyewah et al., 2018; Okyere et al., 2021).

This pathological view of the slum–environmental sustainability nexus is, however, increasingly being challenged by evidence of the regenerative and sustainability potentials of slums and informal settlements generally (Venter et al., 2019; Okyere et al., 2021). Indeed, some sustainable transition activists and scholars have hinted that circularity exists in slums and low-income settings generally (Wright et al., 2019; Marino, 2021). This notion underlies the *'everything in slums is circular'* (see, e.g., Marino, 2021) and *'lower-income countries are more 'circular' than their developed-economy counterparts'* mantras (Wright et al., 2019). Currently, however, the CE practices and processes perceived to underpin slums remain poorly theorized, with the few existing subaltern CE studies exploring such local political economies narrowly focused on waste picking (e.g., Gutberlet et al., 2017; Wuyts and Marin, 2022).

There is an urgent need to document and theorize the slum–circular economy nexus more systematically and holistically. Abunyewah et al. (2023) suggest situating this nexus with three core elements of slums—livelihoods, housing, and spaces (natural and built space) and explicating their synergies with CE principles. Yet, these synergies cannot be oversimplified or merely romanticized due to well-documented trade-offs. For example, in India, key but vulnerable actors, such as e-waste recyclers in the circular chain, are known to face severe health risks such as cancers or even skeletal deformities in babies (See Awashti et al., 2016). Similarly, in two of Ghana's largest cities where informal e-waste scavenging is a major livelihood for the urban poor, scholars (e.g., Oteng-Ababio and Grant, 2020) have reported that scavengers work in toxic environments that expose them to infectious diseases. There are also environmental harms to the natural environment as well: informal recycling activities also pollute or clog nearby rivers. In contrast to slums, formally planned areas of cities often have mechanisms to protect vulnerable citizens against the risks associated with domesticating CE.

Clearly, these environmental and public health risks to informal recycling in urban slums imply conflicts with CE social sustainability principles and environmental justice. Yet, we argue that such conflicts and inherent trade-offs valorize the need to situate CE within everyday life-making practices in slums. Balancing synergies and trade-offs is a critical endeavor to provide new pathways for theoretical and policy discussions about how the synergies can be leveraged through multi-scalar processes that recognize and integrate slum-dwelling activities and also reduce inherent conflicts or risks. Indeed, as Moreau et al. (2017) and Schröder et al. (2020) reasoned: CE trade-offs or conflicts should not supplant but must instead revitalize the need for institutional and collaborative rearrangements and supportive policies that promote skills transfer, improve health and labor conditions, and allow local practices to flourish circular dimension. In other words, such synergetic framings can stimulate new possibilities that mitigate the tendency for CE implementation to reproduce existing inequalities and position informal settlements toward a just sustainability transition. This, however, invites researchers and policymakers to flesh out the underlying contradictions and complexities more carefully and systematically. Consequently, this volume brings together some of the emerging scholarship within the slum–CE nexus to develop a series of policy and theoretical insights that can advance the sustainability and circularity agenda in twenty-first-century built environments of the southern cities.

2 Overview of Chapters

This edited book is a product of our interest in Global South cities, sustainability, and the persistent neglect of the agency of the urban poor and marginalized in the policy and praxis of sustainable urban development. While we have explored these interests in several projects in recent years, one particular project laid the foundation for this book. Our perspective essay, "Advancing slum–circular economy model for sustainability transitions in Global South cities," which was published in *Nature Sustainability* (see Abunyewah et al., 2023), ignited the idea to assemble a collection of research works that unpack the synergies between urban slums and circular economy principles in Global South cities where informality constitute the dominant mode of urbanization and urban growth. This was a modest attempt but a bold call, as we are aware that the CE is just beginning to gain momentum in Global South cities, and urban slums have been considered apart from, not part of, the ongoing discourses for promoting circularities in southern urbanisms.

Rightly so, the editorial comments on our perspective essay highlighted the novelty of synthesizing two important issues—urban slums and circular economy—that remain disparate and poorly connected in both international and local discussions on achieving sustainable development goals in informal cities. For this, we consider the collection of essays in this volume as an entry point for scholars and practitioners to take the urban slum–CE synergies more seriously to situate and offer pathways for sustainable transitions that are equitable, inclusive, and environmentally conscious of local practices and everyday lived experiences. Before delving into the structure and content of the chapters, an important clarification needs to be made. The chapters in this book use urban slums and informal settlements interchangeably, though we understand that while all slums are informal settlements, not all informal settlements are slums (see Okyere and Kita, 2015; Okyere et al., 2021). Therefore, informal settlements are used in this book to include slum settlements that feature poor housing conditions, fragile livelihoods, underserved and with limited access to basic infrastructure services such as water, sanitation, and exposure to disaster risks (Abunyewah et al., 2018). This excludes high-quality unplanned settlements or illegally built or modified housing by the wealthy class that often gets conflated in the nomenclature of informal settlements.

Nonetheless, the use of slums in the title is intentional. It is well-established that international development organizations and policy agencies continue to use the terminology of slums in global cities reports and planning policies. The UN-Habitat Slum Almanac and the UN Sustainable Development Goals Annual Reports all use slums in reporting trends and progress about slum populations and conditions. Thus, the choice of slums in the title allows us to bring the contributions of this book closer to the key actors who hold purse strings and influence in the policy arena where development and planning decisions about slum improvement often take shape.

The book is structured into eleven chapters. Following this introductory chapter, Fianoo and his colleagues, in Chapter Two, present a review "linking informal settler practices and circular principles in Global South: Lessons from Asian and Latin American Cities." They provide a catalog of informal settler practices and how they align with the circular economy's core principles, such as reducing, repurposing, and recycling. Their essay documents the role of informal waste pickers in Belo Horizonte, one of Brazil's largest cities and where large tons of waste are generated annually. What is noticeable in this case is the institutionally supportive and collaborative platform: Municipal Waste and Citizenship Forum, which has brought together civil society organizations, waste picker cooperatives, and the municipal authority to develop a plan that improves the collection of waste and, most importantly, recognizes the role of informal waste pickers through the 'recycling bonus' initiative. This chapter highlights the importance of grassroots organizing and a supportive institutional regime to harness the role of informal waste-dependent livelihoods to both the circular economy and the well-being of waste pickers.

In Chapter Three, award-winning Danish Architect and Researcher Johan Motelsson takes up the theme of informal housing, investigating "modern vernacular architecture and the circular economy." He carefully analyzes the materialities of housing construction in selected cases from South Africa, Asia, and Latin America. In Khâm Thiên, an informal settlement in Hanoi (Vietnam), for example, Mottelson draws attention to material upcycling as observed in the use of "plastic boards or heavy vinyl tarp, providing durable, flexible, waterproof materials for wall cladding" (See Mottelson in this volume). This is a clear example of the repurposing principle in circular economy and thus demonstrates how material recompositions in slum housing constructions can meet the dual needs of shelter and waste reduction

Miller et al. (this volume) explore "Circular Economy Awareness, Perceptions, and Practices in Selected Urban Slums in Kigali City of Rwanda" in Chap. 4. Their field study, geographically situated across three communities, shows low awareness of the term 'circular economy' but widespread livelihood practices that intertwine CE principles such as repurposing or recycling waste materials. An interesting point in their study is the respondents' preference for the *circular* but not the *economy*, as they associate the *economy* with wealth and power in the government and institutional regime. This is instructive and foregrounds this book's emphasis on making circular cities agenda sensitive to the everyday lived realities of the urban poor, especially slum dwellers.

Does construction and demolition (C & D) waste offer any scope for circular management in Dhaka's slum housing sector? This is an important question that is taken up in Chapter Five by Fatemi and Rahman. For the researchers, Dhaka's rapid but poorly regulated urbanization and associated construction boom has contributed to the massive amount of construction and demolition waste, which mostly ends up in landfills. Fortunately, their content analysis of plans and policy documents shows institutional awareness and support for recycling construction and demolition waste in Bangladesh and Dhaka. From a policy perspective, this opportunity could be leveraged to address the challenges around adequate housing for slum populations. Yet, their study found that the preferences of slum dwellers in improving housing conditions from recycled waste are not straightforward. Two-thirds of slum dwellers interviewed in their study were aware of C & D waste, and one-third had used them previously. Their insightful and timely study highlights that recycling C & D waste is not a wholesale adventure, as the cost, quality, and durability of recycled material are important factors to slum dwellers in using them.

Housing is pivotal in the urban experiences of slum dwellers, and their transformations represent the mainstay of getting by in the saturated housing market that seems to cater to the preferences of the wealthy rather than the survival of the poor. In Chapter Six, Avogo et al. (this volume) offer a synergetic review of circularities in housing transformation practices in Global South cities. Their chapter takes a deep dive into the literature on housing transformation to uncover the presence or absence of circular economy principles—weaving together case reviews from settlements in Santiago (Chile) and Nairobi (Kenya) to assert that repurposing waste as building materials to aid housing modifications constitutes the dominant presence of circular economy in transformation practices. Their review and conceptual reflections find empirical validation in Chap. 3, where Mottelson documents how repurposing manifests in the vernacular architecture of urban slums in Vietnam and Mozambique.

In Chapter Seven, Mensah et al. (this volume) present a narrative "review of residents' value retention practices and their implications for participatory urban planning." They categorize urban slum residents' initiatives into urban agriculture, green space development, housing, and transportation. One transformative case in their chapter is how the residents of Nairobi's Dandora slum have converted part of a nearby dumpsite into a community garden (growing food crops) using compost and constructing an informal green space by repurposing and reusing waste materials (e.g., plastic, wood) for outdoor recreation and social gathering. This initiative highlights how waste and spatial appropriations in the slum living environment can intersect to help residents address their socio-environmental deficiencies in somewhat circular ways. The authors conclude that recognizing such initiatives is integral to participatory urban planning in African cities.

Energy poverty is a critical issue for many households living in peri-urban Uganda. In Chap. 8, Natalya Apanovich proposes the transformation of agricultural residues, which remain largely underutilized, into electricity to help meet the energy needs of smallholder farmers in central Uganda. She cautions that while wasted agricultural residues can meet 14 percent of current energy demand using gasification, it cannot be considered circular due to socioeconomic realities (e.g., small-scale subsistence farming, land fragmentation, and land tenure challenges) and the risk of food insecurity and class tensions. Instead of peri-urban areas, the researcher suggests that the potential for using agricultural residue for energy generation lies in urban areas if underpinned by principles such as preserving and conserving natural capital, optimizing resource yields, and system effectiveness of conversion technologies.

In Chapter Nine, Contin and Pandolfi reflect on their 'novel approach for assessing how new morpho-type solutions impact the circular city agenda—the Ouagadougou Case Study.' Drawing on the MSLab research project at the Polytechnic University of Milan (Politecnico di Milano), the chapter argues that participatory-oriented approaches to mapping spatial structures can serve as a tool to manage scarce resources in informal cities such as Ouagadougou. Indeed, collective mapping initiatives are essential to identifying local resources, their use, conditions, and management. Natural and built resources (e.g., green infrastructure) are the core of the circular and sustainable cities agenda, and adopting innovative tools to map such elements in the built environment represents an important step in defining and regulating their use through circular principles.

Chapter Ten presents findings from a narrative review of 'Policy-implementation gaps in e-waste management: Implications for circular economy and sustainability in urban Ghana' conducted by Frimpong, Mensah, and Okyere. Cognizant that e-waste handling poses a health threat to informal e-waste recyclers and is detrimental to the local environment, they employ content analysis to investigate the articulation of circular principles in policies for managing e-waste. They conclude that circular economy principles remain to be functionally integrated into regulation and management policies such as the Hazardous and Electronic Waste Control and Management (HEWCM) Act 2016 (Act 917) and Hazardous, Electronic, and Other Wastes (Classification), Control and Management Regulations (HEWCMR). The chapter highlights that championing the role of informal e-waste recyclers should be accompanied by private (e.g., manufacturers) and public sector (e.g., environmental agencies) partnerships to support skills improvement and capacity building for handling e-waste, safety protections, and livelihood support.

Taken together, the contributions in the edited volume geographically and spatially assemble empirical and review cases from Sub-Saharan Africa, South Asia, and Latin America. Scholars and practitioners encompassing architects, urban planners, geographers, sustainability scientists, environmental planners, and development policy experts have shed insights into everyday practices in urban slums or informal settlements and their synergies with the core principles of the circular economy. As such, the chapters offer useful learnings on subaltern practices and the potential for localization of the circular economy agenda in Global South cities. In contexts where the agency of informal or slum residents is mostly subjugated, obfuscated, and denied

(Okyere et al., 2022; Azunre and Boateng, 2023), this book offers a grounded view of the subaltern actors, practices, and initiatives in advancing the slum–circular economy model for sustainable urban development in Global South cities. It also confronts the conventional perspective of slum settlements as a '*habitus of environmental pathologies*' and contests that other views and understandings require more attention and engagement to foster progress toward sustainable and circular city goals in twenty-first-century urbanism.

References

Abunyewah M, Gajendran T, Maund K (2018) Profiling informal settlements for disaster risks. Procedia Eng 212:238–245

Abunyewah M, Erdiaw-Kwasie M, Okyere S, Boateng FG (2023) Advancing a slum-circular economy model for sustainability transition in global South cities. Nat Sustain

Allam Z, Jones DS (2018) Towards a circular economy: a case study of waste conversion into housing units in Cotonou, Benin. Urban Sci 2(4):118

Awasthi AK, Zeng X, Li J (2016) Environmental pollution of electronic waste recycling in India: a critical review. Environ Pollut 211:259–270

Azunre GA, Boateng FG (2023) Symbolic legitimization of evictions in post-colonial African cities: a review from Ghana's old fadama. Habitat Int 135:102809

Blomsma F, Brennan G (2017) The emergence of circular economy: a new framing around prolonging resource productivity. J Ind Ecol 21:603–614

Bob-Millar GM, Obeng-Odoom F (2011) The informal economy is an employer, a nuisance, and a goldmine: multiple representations of and responses to informality in Accra, Ghana. Urban Anthropol Stud Cult Syst World Econ Dev, 263–284

Bocken NM, De Pauw I, Bakker C, Van Der Grinten B (2016) Product design and business model strategies for a circular economy. J Ind Prod Eng 33(5):308–320

Christensen TB (2021) Towards a circular economy in cities: exploring local modes of governance in the transition towards a circular economy in construction and textile recycling. J Clean Prod 305:127058

Corvellec H, Stowell AF, Johansson N (2022) Critiques of the circular economy. J Ind Ecol 26:421–432

Davis M (2013) Planet of slums. New Perspect Q 30(4):11–12

D'Odorico P, Davis KF, Rosa L, Carr JA, Chiarelli D, Dell'Angelo J, Rulli MC (2018) The global food-energy-water nexus. Rev Geophys 56(3):456–531

Ellen McArthur Foundation (EMF) (2013) Towards the circular economy: Economic and business rationale for an accelerated transition

Engstrom R, Pavelesku D, Tanaka T, Wambile A (2019) Mapping poverty and slums using multiple methodologies in Accra, Ghana. In: 2019 joint urban remote sensing event (JURSE)

Geissdoerfer M, Savaget P, Bocken NM, Hultink EJ (2017) The circular economy—a new sustainability paradigm? J Clean Prod 143:757–768

Gutberlet J, Uddin SMN (2017) Household waste and health risks affecting waste pickers and the environment in low- and middle-income countries. Int J Occup Environ Health 23(4): 299–310.

Kębłowski W, Lambert D, Bassens D (2020) Circular economy and the city: an urban political economy agenda. Cult Organ 26(2):142–158

Klopp JM, Paller JW (2019) Slum politics in Africa. In: Oxford research encyclopedia of politics

Lüdeke-Freund F, Gold S, Bocken NM (2019) A review and typology of circular economy business model patterns. J Ind Ecol 23(1):36–61

Mahabir R, Crooks A, Croitoru A, Agouris P (2016) The study of slums as social and physical constructs: challenges and emerging research opportunities. Reg Stud Reg Sci 3:399–419

Marino G (2021) The slum economy: an alliance between the circular and the informal. Renew Matter. https://www.renewablematter.eu/articles/article/the-slum-economy-an-alliance-between-circular-and-informal

Moreau V, Sahakian M, van Griethuysen P, Vuille F (2017) Coming full circle: why social and institutional dimensions matter for the circular economy. J Ind Ecol 21(3):497–506

Obeng-Odoom F (2011) The informal sector in Ghana under siege. J Dev Soc 27(3–4):355–392

Obeng-Odoom F (2013) Governance for pro-poor urban development: lessons from Ghana. Routledge, NY

Okyere AS, Kita M (2015) Rethinking urban informality and informal settlements growth in urban Africa: a literature discussion. J Sustain Dev Afr 17(2):101–124

Okyere SA, Diko SK, Ebashi M, Kita M (2022) Obscured innovations? Inventiveness in collective infrastructure management in Accra, Ghana. Informality and the City: Theories, Actions and Interventions. Springer International Publishing, Cham, pp 555–568

Okyere SA, Kusi LF, Diko SK, Abunyewah M, Kita M (2021) Situating everyday urban struggles within the context of the SDGs in an informal settlement in Accra. In: Cobbinah P, Addaney M (eds) Sustainable urban futures in Africa. Routledge, New York/London

Oteng-Ababio M, Grant R (2020) E-waste recycling slum in the heart of Accra, Ghana: the dirty secrets. In: Handbook of electronic waste management. Butterworth-Heinemann, pp 355–376

Paller JW (2015) Informal networks and access to power to obtain housing in urban slums in Ghana. Africa Today 62(1):31–55

Paller JW (2019) Democracy in Ghana: everyday politics in urban Africa. University Press, Cambridge

Paller JW (2014) Informal institutions and personal rule in urban Ghana. Afr Stud Rev 57(3):123–960 142

Patel J (2023) Circular economic approach to Dharavi: existing self-build framework intervention.

Schröder P, Lemille A, Desmond P (2020) Making the circular economy work for human development. Resour Conserv Recycl 156:104686

Schwanholz J, Leipold S (2020) Sharing for a circular economy? An analysis of digital sharing platforms' principles and business models. J Clean Prod 269:122327

Stahel WR (2016) The circular economy. Nature 531(7595):435–438. https://doi.org/10.1038/531435a

Turok I, Borel-Saladin J (2018) The theory and reality of urban slums: pathways-out-of-poverty or cul de-sacs? Urban Stud 55(4):767–789

UN-Habitat (2008) State of the world's cities 2010/2011: bridging the urban divide. Earthscan Publications Ltd., London

UN-Habitat (2016) Slum almanac 2015–2016: tracking improvement in the lives of slum dwellers. UN-Habitat, Nairobi

UN-Habitat (2020) World cities report 2020: the value of sustainable urbanization. https://unhabitat.org/wcr/. Accessed 10 March 2022

Venter A, Marais L, Morgan H (2019) Informal settlement upgrading in South Africa: a preliminary regenerative perspective. Sustainability 11:1–15

Williams J (2019) Circular cities. Urban Stud 56:2746–2762

Wright CY et al (2019) Circular economy and environmental health in low-and middle-income countries. Global Health 15:65

Wuyts W, Marin J (2022) "Nobody" matters in circular landscapes. Local Environ 1–18.

Linking Informal Settler Practices and Circular Principles in Global South: Lessons from Asian and Latin American Cities

Isaac Nevis Fianoo, Stephen Leonard Mensah, Seth Asare Okyere, Matthew Boadie-Ampong, and Joshua Jirjiri

Abstract Informal urbanization in the Global South has, over the years, received scholarly and policy attention. Amidst this sphere of urbanism are the growing socioeconomic and environmental laggardness and challenges which drive the adoption of circularity practices. This chapter aims to examine the linkages between informal settler practices and circularity principles and how they align with the sustainable cities agenda. It draws on a review of secondary sources such as journal articles, grey literature, and official government reports to present evidence from India, Brazil, Bangladesh, and Argentina. The findings reveal that informal settlers engage in solid waste recycling for building purposes. It was also revealed that leftover pieces of wood and wooden pallets are reused for furniture and home decor. This chapter, therefore, recommends support from governments and the private sector for the adoption of adaptable circular practices evident in informal settlements, which have generally been regarded as the origin of most urban problems.

Keywords Global south · Informal settlers · Circular practices · Sustainability · Circular economy

I. N. Fianoo (✉) · M. Boadie-Ampong
Institute for the Advanced Study of Sustainability, United Nations University, Tokyo, Japan
e-mail: fianoo@student.unu.edu

S. L. Mensah
Department of City and Regional Planning, The University of Memphis, Memphis, TN, USA

S. A. Okyere
School of Landscape Architecture and Planning, University of Arizona, Tucson, AZ, USA
e-mail: sethaokyere@arizona.edu

J. Jirjiri
Department of Development Management, SD Dombo University of Business and Integrated Development Studies, Wa, Ghana

S. A. Okyere et al. (eds.), *Urban Slums and Circular Economy Synergies in the Global South*, Advances in 21st Century Human Settlements,
https://doi.org/10.1007/978-981-99-9025-2_2

1 Introduction

Over 56% of the global population resides in cities (World Bank 2022), and this is expected to grow exponentially in the next decades. Rapid urbanization creates numerous opportunities such as large markets for raw materials and stimulates opportunities for local economic development. Yet, most urban south cities are characterized by complex urban problems such as poor waste management practices. Policy and planning responses to these problems have largely been unsuccessful due to its exclusionary approach (Obeng-Odoom 2013) to urban policy planning. Akin to these is the spatial manifestation of informal urbanization, i.e., informal settlements that have become urban centers for experimental initiatives which align with circularity practices. According to Martínez et al. (2018), informal settlers contribute substantially to the local urban economy. Yet, their shelter and activities have been considered unlawful and detrimental to urban sustainability.

Globally, approximately 23.5% of people live in informal settlements (UN Statistical Division 2021). The proportion of informal settlers has increased to over 1 billion with 80% residing in Global South (GS) regions such as Eastern and South-Eastern Asia (370 million), sub-Saharan Africa (238 million), Central and Southern Asia (227 million), and 20% in other regions (UN Statistical Division 2021). According to the UN-Habitat (2018a, b), one out of eight people globally live in a settlement with the characteristics of informal settlements. The population of residents in informal settlements across the globe is expected to increase by 2025 (see Fig. 1).

Informal settlement remains the most common form of spatial representation of urban settlement in the world (Dovey et al. 2020). But the inability to make space for their needs in city planning responses has resulted in deteriorating environmental conditions. For instance, they face inadequate and overburdened infrastructures such

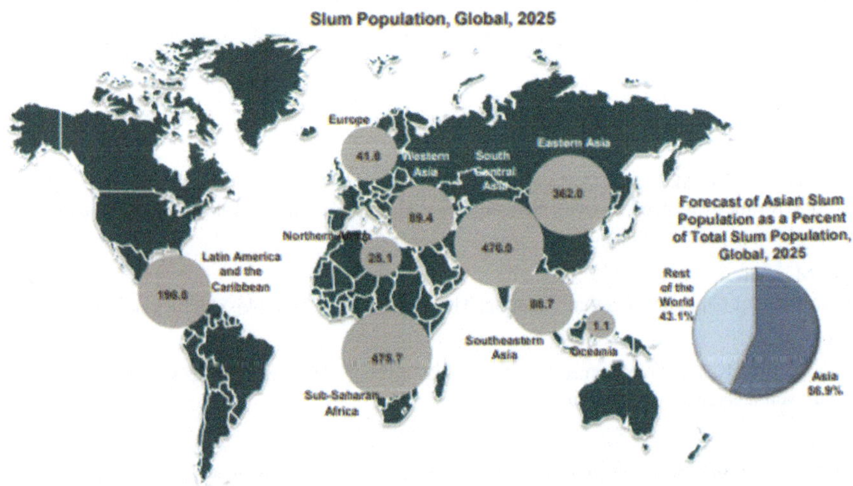

Fig. 1 Projections of Slum Population by 2025 (*Source* Adapted from the UN-Habitat 2018a, b)

as poor water services, sanitation facilities, housing durability, and insufficient living space (UN-Habitat 2021). These socioeconomic and environmental challenges have given rise to concern for urban sustainability, especially toward the achievement of SDG 11. Achieving sustainable urban development requires innovative and creative living. Here, informal settlers provide ample opportunities and lessons for city authorities as much of their activities and livelihoods are embedded in circular practices. Against this backdrop, this chapter reviews the everyday activities of informal settlers in the selected countries of Argentina, Bangladesh, Brazil, and India and inductively draws sustainable lessons for urban circularity and sustainability. This chapter is significant because the circular economy (CE) concept has become an alternative model to the linear economy in urban development and planning; hence lessons from grassroots and urban spaces regarded as 'illegal' are important for knowledge co-production and equitable urban development. The next section provides a conceptual understanding of the nexus between informal urbanization and CE. This is followed by the methodology and the description of case studies. The next is the discussion of the lessons and the conclusion.

2 Informal Urbanization and Circular Economy: Opportunities and Challenges

Informal settlement, also known as a slum, shantytown, or squatter settlement, is an area where people live in inadequate housing conditions without secure land tenure or basic services such as water, sanitation, and electricity (Okyere and Kita 2015; Nuissl and Heinrichs 2013), growing worldwide in tandem with increasing urban population. Since 1990, it has been estimated that 213 million people, or around 25% of the global urban population, have lived in informal settlements (UN-Habitat 2018a, b). These are residential areas where (Brown and McGranahan 2016; UN-Habitat 2011) people frequently lack security of tenure for the land or dwellings they occupy; neighborhoods typically lack basic services; city infrastructure or housing may not comply with planning and building regulations; and is frequently located in geographically and environmentally sensitive areas. Population growth, rural–urban migration, a lack of affordable housing, and poor governance (especially in policy, planning, and urban management), are some of the interconnected factors that have contributed to their emergence (UN-Habitat 2022, 2018a, b).

Sean Fox (2014), further highlighted the restriction of access to urban space from the colonial institutions of urban governance, including tenancy rights plot allocation, land registration procedures, and building codes as the other factors to the emergence of informal settlements. They are also characterized by poor socio-environmental and health conditions (Brown-Luthango et al. 2017). Most city authorities have found it a daunting task planning for informal settlements as their status in planning provisions are often stipulated as illegal spaces (Satterthwaite 2017; Mwije 2013). Informal settlements occupy about 90% of the land in developing countries and have received heightened scholarly attention (Duranton and Puga 2014; UN-Habitat 2018a, b).

As the global population of slum dwellers is expected to increase by 2025 (see Fig. 1) (Todes 2011), informal settlements are expected to increase in Global South cities (Sminkey and Le Doux 2016). Different informal settlements are underpinned by different causative factors as some are 'legal but unplanned' (Dovey and King 2011). Concerns toward achieving sustainable urban development amidst the increasing rise in informal settlements requires alternative and inclusive urban development models (Muchangos 2022) over the existing linear urban development models. Here, the CE model presents opportunities and lessons embedded in the everyday activities of informal settlers for city authorities. The CE concept emerged because of the need for a sustainable economic system. Notwithstanding its promise of a green economy, it failed to address the amenity base and life-support features of the surrounding environment (Inigo and Blok 2019). It also addresses neither the "critical importance of land as the basic source of biomass, energy, and mineral reserves" (Winans et al. 2017, p. 829). Amidst all the criticisms, circular practices have become crucial due to the inherent challenges associated with the linear model. It prioritizes practices of minimizing resource intake, use, and emissions while also repairing, reusing, and recycling resources, aiming to preserve materials at their best intrinsic value for as long as feasible. By so doing waste reduction is achieved and harmful exigencies such as CO_2 emissions are minimized (Ingemarsdotter et al. 2019). Resource utilization has increased due to rapid population growth and income levels (Behrens et al. 2007). Consequently, it is expected that as resources become scarce and expensive, circular practices would be adopted to meet everyday socioeconomic needs (Chigumira 2016). These practices may be seen in the design of informal spaces, shelters, the use of materials, and the way residents engage each other. Lessons from these practices and collaborative planning responses are crucial for building sustainable and resilient circular cities in the Global South (Garam et al. 2019). Growing societal, family, and personal needs requires circular practices to meet socioeconomic needs and reduce household expenditure (Quesada-Román 2022).

Informal settlements are not just places of run-down houses and poor environmental conditions but urban places with a strong sense of attachment and community with innovative and creative experimental ideas (Allam and Jones 2018). They are places that manifest circular practices, and these are evident in their architecture, construction of homes, the materials used, and residents' human–environment interactions. Lessons drawn from these practices are crucial for climate mitigation and adaptation strategies and open opportunities for poor urban populations (Cities 2022). Incorporating lessons from such practices into formal planning and policy responses can enhance their effectiveness and help build inclusive and resilient cities (Muchangos 2022).

Lessons from circular practices are crucial in addressing environmental problems such as pollution, waste, biodiversity loss, and climate change. It is also useful for resource efficiency, low-carbon investment, and the sustainability of the growing production and consumption systems across the world. It is based on the design principles of reducing waste and pollution, reusing products and materials, and restoring the natural environment (Akhimi et al. 2020). Formalizing circular practices has become daunting due to low levels of awareness, financial challenges, low levels of technology, and the lack of effective collaboration between the informal and the formal players (Ghisellini et al. 2016).

3 Profiles of the Setting of the Study

This section of the chapter presents informal settlement practices from the selected countries form the Global South, such as India, Brazil, Bangladesh, and Argentina. Table 1 shows selected cases from informal settlements within these countries and their associated circularity practice. The scholarly secondary literature used was purposively selected to illustrate the potential of informal settlements to support building CE cities in the global south. This literature included journal articles, websites (e.g., Sustainable Recycling Industries), and reports from private organizations.

3.1 Reuse

3.1.1 The Reuse of Waste Materials for Housing Development

Dharavi is one of the well-known informal settlements in India and houses over 1 million people, it is one of the most densely populated slums in the world, with a population density of approximately 277,136 km^2 (Basu 2022). The buildings in the community were built, having both shops and business enterprises on the

Table 1 Selected cases of informal settlements practices

Pillars of circularity	Practice description
Reuse	The use of waste materials for housing development, Dharavi, India
Recycle	Waste Concern, recycling and Waste management, Kalyanpur, Bangladesh Slum-based businesses recycle cardboards and aluminum waste, Belo Horizonte, Brazil
Repurpose	The use of plastic bags as storage facilities, Rocinha, Brazil
Reduce	Waste picking to reduce waste in Calle Paraguay, Palermo Soho, Argentina Improving housing conditions to reduce climatic hazards, Korail, Bangladesh Vegetable gardens for provisional purposes, Barrio San Jorge, Argentina Community-designed toilet initiatives (Patel and The SPARC Team 2015), Pune, Mumbai, Bangalore, Kanpur, India

ground floor, and at night, the structures on the upper floors are used as cubicles for families to sleep. It has a housing unit of about 20 m^2 and shelters an average of 10 people. Likewise, in most informal settlements, sanitation infrastructure is typically inadequate, and inhabitants usually depend on water tankers to meet their domestic and drinking water needs (Basu 2022). Information on waste management and its flow in the community is mostly informal (Dey 2018). Residents also face dire social exclusion resulting in a struggle with adequate housing infrastructure and the lack of essential social amenities (Basu 2022). Housing deficit is a major problem in the community, resulting in overcrowding and large room occupancy. This can lead to the rapid spread of communicable and infectious diseases. In response to their housing challenges and commitment to the circularity principles, residents decided to use materials such as wooden sticks, bamboo, and corrugated metal sheets to make their own housing units. These housing structures begin with embedding wooden sticks or bamboo into the ground, which then acts as a frame to support the structure for a metal or plastic sheet that serves as a wall. Roofing materials for these buildings are usually made from old and discarded metal containers or drums, which they stretch to create a flat metal sheet (Dey 2018).

3.2 *Recycle*

3.2.1 Slum-Based Enterprises (Cardboards and Aluminum Waste Recycling)

Belo Horizonte is one of the largest cities in Brazil, with a population of over 2.5 million people (Caldas 2019). The city has several informal settlements totaling about 324 (Caldas 2019), which creates a large amount of waste. Evidence suggests that the informal sector contributes substantially to recycling in Brazil, 92% for aluminum and 80% for cardboard, respectively (Dias 2016). As a result, city authorities in 1993 initiated a program that integrated them into the waste management system to control waste generation and adhere to the circularity principles. The program assembled into two categories of waste pickers; those who collect recyclables and debris. In 2003, city authorities also created a forum formed by Civil Society Organizations, workers within the waste management system, and internal staff called the Municipal Waste and Citizenship Forum (Dias 2014).

The forum developed a comprehensive plan which promoted waste segregation at the source, where waste pickers were key players in the collection and sorting of waste. The plan comprised three main pathways; curbside recycling collection, "drop-off points" where people can store recyclables, and door-to-door recycling collection in specified areas. Cooperatives within the waste management system often use vehicles or carts to perform door-to-door gathering of waste. After being gathered through drop-off methods, recyclables are processed and sorted by informal recyclers. Informal workers admit that the city's approach to integration has improved their lives over the past 20 years (Korsunova et al. 2022). In recognition of the work done by waste pickers, the state government instituted an incentive package called the Recycling Bonus. The amount of money to be received depends on the type and quantity of recyclables gathered and sold, which would be distributed to individual members of the cooperatives (Dias 2016).

3.2.2 Waste Concern, Recycling, and Management

Kallyanpur is an informal settlement in Dhaka, Bangladesh. A few studies have documented how informal settlers within this settlement have resorted to adopting innovative ways to meet their socioeconomic needs in the city as well as manage the rate of waste production and disposal (Parvin et al. 2016; Roy 2011). Slum dwellers are said to provide high-quality, affordable domestic and retail services to non-slum areas, including recycling services. To find recyclables, slum dwellers labor as waste pickers who scour the streets, trash cans, transfer stations, and dumps in the surrounding Kallyanpur region and adjoining settlements. As a result, this helps the larger city by lowering its environmental impact, which would otherwise deteriorate because the local institution in charge of these services is under-resourced and unable to handle the sheer volume of Dhaka's waste (Waliuzzaman and Alam 2022). Residents within this informal settlement engage in waste sorting, cleaning, and selling of waste materials to waste firms. In the event of rain, they dry picked waste for a day or two before selling them to recycling companies.

3.3 *Repurpose*

3.3.1 The Use of Plastic Bags as Storage Facilities

The amount of waste humans produce from households is growing worldwide, and its composition is getting more complicated. City managers are charged with safeguarding their citizenry from waste implication and conformity to waste management principles, but due to challenges such as financial and inadequate logistics, these duties are mostly not achieved. However, efforts are made by communities to rely on waste pickers, waste micro-entrepreneurs, and recycling groups to collect and recycle waste (Gutberlet and Carenzo 2020). Beyond the contributions rendered by these actors, informal settlers also play a major role. In Rocinha, Brazil, informal settlers found a way to deal with plastic waste. They repurpose these plastic bags and use them as storage bags. For example, to prevent items used within their home from being found on bare floors. Due to the lack of space within shantytowns for furniture sets, shelves, and drawers, essential papers, and other items are placed on the wall in plastic bags, including toothbrushes and other personal hygiene supplies. Children within informal settlements have used lessons from schools on how to reuse Polyethylene terephthalate (PET) bottles as plant pots, for instance, which helps to stop the spread of dengue disease by limiting insects' access to still water (Korsunova et al. 2022).

3.4 Reduce

3.4.1 Barrio Vegetable and Kitchen Gardens for Provisional Purposes

Barrio San Jorge is a low-income settlement within the Buenos Aires Metropolitan Area, Argentina, and home to about 400 families (Hardoy et al. 2010). Within this settlement, the youth group is the most vulnerable as they tend to fight stigma from adults in the community and their colleagues from rich neighborhoods. In their study conducted, Hardoy et al. (2010) revealed that most youths drop out of school and end up without any formal education within the informal sector. Girls between the ages of 17 and 19 assume early household duties and become teenage mothers. These circumstances restrict their ability to advance their capabilities, acquire new skills, and escape informality. To help the youths within the community, a project was initiated to train the youth in basic gardening skills, which aimed at equipping them with the needed skills that could be useful for potential future employment opportunities. As a result, vegetables were cultivated for local consumption, and some sold out on the market. Groups were formed involving friends or relatives, and they were tasked to develop a work schedule, design the vegetable garden, plant, and maintain the garden. One of the major challenges of this project is the limited financial resources and encouraging the use and knowledge of organic products. Also, due to the social divide between adjourning neighborhoods, vegetables proved very challenging to sell. However, the project was recommended in 2009 and is presently operating without funding, with a focus on three distinct groups, children, youths, and adults (Hardoy et al. 2010). This project has the potential to reduce food insecurity within the city.

3.4.2 Waste Picking to Reduce Waste

Calle Paraguay in Palermo is an informal settlement also within the Buenos Aires Metropolitan Area, Argentina. Residents (men, women, and children) within this settlement engage in litter picking to reduce the waste in their neighborhood. The items they pick range from discarded cardboard, plastic bottles, discarded check-books, set of bookends, and newspapers. These wastes, when collected, are packed in large sacks and kept by the roadside, then trucks come to pick them up for recycling. Attempt to promote recycling led to city authorities enacting a law in 2005 known as Zero Garbage Law which aimed to reduce gradually the volume of waste that goes to the landfill. Waste pickers were also given monthly incentives for emptying waste into recycling bins within the settlement. They were also placed on a social security package and a pension scheme (Balch 2016). These acts by residents help to improve the environmental quality of the city, thereby promoting sustainability.

3.4.3 Improving Housing Conditions to Reduce the Impacts of Climate Hazards

Korail is the largest informal settlement in Dhaka, Bangladesh, and has a population of at least 100,000 people (Sinthia 2020). The slum has a high population density and lacks essential social amenities and infrastructure. Part of its boundaries falls near Lake Gulshan, which makes it susceptible to the dangers brought on by both climate change and variability. Like most slums, Korail is a settlement with low-income earners. Residents of the settlement annually experience weather hazards like flooding, extreme heat, and excessive rainfall. To deal with the phenomenon of extreme heat, leafy vines are grown to cover the roofs of the buildings, which are constructed of corrugated metal sheets to reduce heat. Again, most households use cloth as a canopy or artificial ceiling to reduce the impact of extreme heat (Jabeen et al. 2010).

3.4.4 Community-Designed Toilet Initiatives

The Indian Alliance Project began in 1993, when a group of Mumbai pavement dweller women built a community toilet which was completed within weeks. It included a water tank, four pour-flush toilets, and a caretaker's room. The Alliance comprised the National Slum Dwellers Federation (NSDF), the women savings group (Mahila Milan), and the Society for the Promotion of Area Resource Centres (SPARC). It allowed the urban poor to influence their policies, programs, and projects (Appadurai 2001; Tomlinson 2015). When their work started, sanitation became a pressing issue in Mumbai. This led NSDF and SPARC, both advocates for sanitation-related issues, to collaborate with these groups to develop secure, long-term housing for themselves. It was apparent that having toilets in their homes was unfeasible due to space and financial constraints, but the need for toilets was a priority (Patel and The SPARC Team 2015; Burra et al. 2003).

After the Alliance's first community toilet project along the P D'Mello Road, Mumbai, the municipality's approach to toilet provision was more challenging than the Alliance's. This was because they proposed a community-led project with the municipality providing funds since there were complaints about the maintenance of toilet blocks (Patel and The SPARC Team 2015). This was the first time the municipality commissioned residents from low-income neighborhoods to build a toilet facility for specific slum communities. Together with experts recommended by SPARC, locals from the area contributed to the design of the toilet blocks and provided unskilled labor from among community members and Mahila Milan.

The Indian Alliance's strategy for sanitary improvement indirectly addressed the four primary issues. They provided some innovations with communal block latrines (i.e., prioritizing affordability), secured funding from the government (i.e., developing a co-productive relationship), placed community members and local organizations at the center of the initiative (i.e., addressing the collective action challenge), and used toilet construction in tandem with other development efforts to change the negative perception of informal settlements and their residents (intended to improve housing security) (Mcgranahan and Mitlin 2016). The initiative also promoted five fundamental activities: community organization and capacity building, precedent-setting, knowledge exchange, partnerships, and policy advocacy (Tomlinson 2015). Despite the importance of the project in dealing with matters related to sanitation issues, unfortunately not all slum settlements have such projects.

4 Implications for Environmental Sustainability and Climate Resilience

In recent years informal settlements have become important places of creative, inventive, and innovative experimental practices with relevance for the circular economy. For instance, informal settlers have engaged in creative urban gardening practices, waste management, and urban mining. These activities reduce the volume of environmental waste and provide opportunities for climate adaptation and resilience. Lessons from these practices are crucial to city authorities to develop strategies for building urban climate resilience and reducing environmental waste. The findings from the review showed that activities of informal dwellers, such as vegetable gardens (Hardoy et al. 2010), while providing food for local and external consumption, also regulate the urban climate. Rapid urbanization has resulted in the depletion of urban forests and the degradation of urban ecologically sensitive areas. Considering that most informal settlements are located close to ecologically sensitive areas in cities, activities that promote urban ecological integrity need to be supported. Urban gardens provide employment to most people leading to improvement in livelihoods (Hardoy et al. 2010). Here, city authorities could tap into this creative initiative to reduce unemployment and achieve urban food security.

Lessons from the review also show that various activities, including waste recycling, reuse, and reduction, are critical in achieving environmental sustainability. Non-degradable waste products such as plastics pose numerous serious risks to fauna and flora (Korsunova et al. 2022). For instance, most animals have lost their lives because of feeding on plastics. Plastics also lead to poor soil filtration, thereby depriving plants of essential nutrients. Having said this, any efforts coming from informal settlers targeted at reducing environmental waste should be commended. The review has shown that informal settlements hold great potential for urban environmental and climate transitions. Despite the numerous challenges that confront them, the invaluable opportunities and lessons they offer cannot be underestimated. Incorporating their practices in formal planning regulations requires collaborative and equitable planning responses from city authorities. It also requires a behavioral change from city authorities as informal settlements are often regarded as places where urban environmental problems emanate (see UN-Habitat 2022, 2018a, b).

5 Conclusion and the Way Forward

Informal settlements have become important urban places noted for creative and innovative practices embedded in circular practices. With CE becoming the alternative model for urban development rather than the linear development model considering the opportunities it presents in preserving the environment and reducing resource consumption. In view of these great potentials, this chapter reviewed experimental informal settler practices in the Global South and drew lessons of urban circularity critical for the sustainability of the urban environment. It was found that experimental practices such as urban gardening as well as waste picking, reuse and recycling are important practices that reduce environmental waste and create opportunities for building climate resilience. These position informal settlements as crucial partners of urban development, which departs from the popular notion that they are the origins of urban environmental problems. Having said this, the future of CE in urban south cities requires collaborative and equitable planning responses, encouragement and support from city authorities, and incorporating residents of informal communities' views in formal planning decisions. Efforts should be made by city authorities to bring solutions and concerns to the challenges faced by these informal settlers. Through these, settlers would feel part of the urban ecosystem and would be motivated enough to initiate creative ideas and experiments for sustainable urban planning. City authorities are also to encourage entrepreneurship and assist local informal enterprises with finances and logistics. These would improve urban mining, an important activity targeted at resource recovery and circularity.

References

Allam Z, Jones D (2018) Towards a circular economy: a case study of waste conversion into housing units in Cotonou, Benin. Urban Sci 2(4):118. https://doi.org/10.3390/urbansci2040118

Appadurai A (2001) Deep democracy: Urban governmentality and the horizon of politics. Environ Urban 13(2):23–43. https://doi.org/10.1177/095624780101300203

Balch O (2016) The new generation of Buenos Aires trash pickers reenergizing recycling in the capital. https://www.theguardian.com/environment/2016/jan/20/buenos-aires-litter-pickers-cartoneros-recycling-argentina-environment

Basu U (2022) Dharavi – a city within a city. https://urbandesignlab.in/dharavi-a-city-within-a-city/ Accessed: March 21, 2023

Behrens TEJ, Woolrich MW, Walton ME, Rushworth MFS (2007) Learning the value of information in an uncertain world. Nat Neurosci 10(9):1214–1221. https://doi.org/10.1038/nn1954

Brown-Luthango M, Reyes E, Gubevu M (2017) Informal settlement upgrading and safety: experiences from Cape Town, South Africa. J Housing Built Environ 32(3):471–493. https://doi.org/10.1007/s10901-016-9523-4

Brown D, McGranahan G (2016) The urban informal economy, local inclusion and achieving a global green transformation. Habitat Int 53:97–105. https://doi.org/10.1016/j.habitatint.2015.11.002
Burra S, Patel S, Kerr T (2003) Community-designed, built and managed toilet blocks in Indian cities. Environ Urban 15(2):11–32. https://doi.org/10.1177/095624780301500202
Caldas MFD (2019) Affordable Housing and Social Inclusion in Belo Horizonte. https://www.thegpsc.org/sites/gpsc/files/maria_caldas_eng.pdf
Chigumira HS (Nov 2016) In-situ informal settlement upgrades through community savings: possibility or farfetched dream in the South African Context
Cities A (2022) Rapid urbanisation and the circular economy, p 6. https://www.citiesalliance.org/newsroom/events/rapid-urbanisation-and-circular-economy
Dey S (2018) The circular economy of Dharavi: making building materials from waste. https://etda.libraries.psu.edu/catalog/15724szd62
Dias SM (Jul 2014) Waste and citizenship forum's trajectories—achievements and challenges. In: International political science association world congress, montreal, Canada, pp 19–24
Dias SM (2016) Waste pickers and cities. Environ Urban 28(2):375–390. https://doi.org/10.1177/0956247816657302
Dovey K, van Oostrum M, Chatterjee I, Shafique T (2020) Towards a morphogenesis of informal settlements. Habitat Int 104:102240
Dovey K, King R (2011) Forms of informality: morphology and visibility of informal settlements. Built Environ 37(1):11–29. https://doi.org/10.2148/benv.37.1.11
Duranton G, Puga D (2014) The growth of cities. In: Handbook of economic growth, vol 2, pp 781–853. https://doi.org/10.1016/B978-0-444-53540-5.00005-7
Fox S (2014) The political economy of slums: theory and evidence from Sub-Saharan Africa. World Dev 54:191–203
Garam B, van Brunschot C, Easen N, Gray V, Kuehr R, Milios A, Iyngararasan M, Pennington J (Jan 2019) A new circular vision for electronics: time for a global reboot. World Econ Forum, 1–24. www.weforum.org
Gethsemane Akhimi N, Latfi E, Shan Shan H (2020) Application of circular economy principles in buildings: a systematic review. J Build Eng 44(m):24–27
Ghisellini P, Cialani C, Ulgiati S (2016) A review on circular economy: the expected transition to a balanced interplay of environmental and economic systems. J Clean Prod 114(May 2017):11–32. https://doi.org/10.1016/j.jclepro.2015.09.007
Gutberlet J, Carenzo S (2020) Waste pickers at the heart of the circular economy: a perspective of inclusive recycling from the global south. Worldw Waste: J Interdiscip Stud 3(1):6. https://doi.org/10.5334/wwwj.50
Hardoy J, Sierra G, Tammarazio A, Ledesma G, Ledesma L, García C (2010) Learning from young people and from our own experiences in Barrio San Jorge. Environ Urban 22(2):371–387. https://doi.org/10.1177/0956247810379964
Ingemarsdotter E, Jamsin E, Kortuem G, Balkenende R (2019) Circular strategies enabled by the internet of things-a framework and analysis of current practice. Sustainability (Switzerland), 11(20). https://doi.org/10.3390/su11205689
Inigo EA, Blok V (2019) Strengthening the socio-ethical foundations of the circular economy: lessons from responsible research and innovation. J Clean Prod 233:280–291.
Jabeen H, Johnson C, Allen A (2010) Built-in resilience: learning from grassroots coping strategies for climate variability. Environ Urban 22(2):415–431. https://doi.org/10.1177/0956247810379937
Korsunova A, Halme M, Kourula A, Levänen J, Lima-Toivanen M (2022) Necessity-driven circular economy in low-income contexts: how informal sector practices retain value for circularity. Glob Environ Chang 76:102573. https://doi.org/10.1016/j.gloenvcha.2022.102573
Martínez L, Short JR, Estrada D (2018) The diversity of the street vending: a case study of street vending in Cali. Cities 79:18–25. https://doi.org/10.1016/j.cities.2018.02.018

Martínez MA, Roitman S (2019) Informal settlers. Wiley Blackwell Encycl Urban Reg Stud 1:1–5. https://doi.org/10.1002/9781118568446.eurs0160

McGranahan G, Mitlin D (2016) Learning from sustained success: how community-driven initiatives to improve urban sanitation can meet the challenges. World Dev 87:307–317. https://doi.org/10.1016/j.worlddev.2016.06.019

dos Muchangos LS (2022) Mapping the circular economy concept and the global south. Circ Econ Sustain 2(1):71–90. https://doi.org/10.1007/s43615-021-00095-0

Mwije S (2013) Corruption and poor governance the major causes of poverty in many third world countries, pp 0–33

Nuissl H, Heinrichs D (2013) Slums: perspectives on the definition, the appraisal and the management of an urban phenomenon. Erde 144(2):105–116. https://doi.org/10.12854/erde-144-8

Obeng-Odoom F (2013) Governance for pro-poor urban development: lessons from Ghana. Routledge. https://doi.org/10.1080/00291951.2014.924552

Okyere AS, Kita M (2015) Rethinking urban informality and informal settlements growth in urban Africa: a literature discussion. J Sustain Dev Afr 17(2):101–124. https://doi.org/10.1177/08854122221128762

Parvin A, Alam AFMA, Asad R (2016) A built environment perspective on adaptation in urban informal settlements, Khulna, Bangladesh. In: Roy M, Cawood S, Hordijk M, Hulme D (eds) Urban poverty and climate change: life in the slums of Asia, Africa and Latin America. Routledge, London and New York, pp 73–91

Patel S, SPARC Team (2015) The 20-year sanitation partnership of Mumbai and the Indian alliance. Environ Urban 27(1):55–72

Quesada-Román A (2022) Disaster risk assessment of informal settlements in the global south. Sustainability (Switzerland) 14(16). https://doi.org/10.3390/su141610261

Roy A (2011) Slumdog cities: rethinking subaltern urbanism. Int J Urban Reg Res 35:223–238. https://doi.org/10.1111/j.1468-2427.2011.01051.x

Satterthwaite D (May 2017) Successful, safe and sustainable cities: towards a New Urban Agenda. Commonw J Local Gov, pp 3–18. https://doi.org/10.5130/cjlg.v0i19.5446

Sinthia SA (2020) Analysis of urban slum: case study of Korail Slum, Dhaka. Int J Urban Civil Eng 14(11):416–430

Sminkey PV, Le Doux J (2016) Case management ethics: high professional standards for health care's interconnected worlds. In: Professional case management, vol 21, Iss 4. https://doi.org/10.1097/NCM.0000000000000166

Todes A (2011) Reinventing planning: critical reflections. Urban Forum 22(2):115–133. https://doi.org/10.1007/s12132-011-9109-x

Tomlinson R (2015) Scalable community-led slum upgrading: the Indian Alliance and community toilet blocks in Pune and Mumbai. Habitat Int 50:160–168. https://doi.org/10.1016/j.habitatint.2015.08.020

UN-Habitat (2018a) County annual progress report, 2018. https://www.makueni.go.ke/site/files/2018/12/Makueni-County-Annual-Progress-Report-2018-1.pdf

UN-Habitat (2018b) SDG indicator 11.1.1 training module: adequate housing and slum upgrading. United Nations Human Settlement Programme (UN-Habitat), Nairobi. https://unhabitat.org/sites/default/files/2020/06/indicator_11.1.1_training_module_adequate_housing_and_slum_upgrading.pdf. Accessed 29 Mar 2023 at 10:34 am

UN-Habitat (2022) Envisaging the future of cities. In: World city report

UN-Habitat (2021) Urban indicator database. https://data.unhabitat.org/pages/housing-slums-and-informal-settlements. Accessed 16 Feb 2023 at 01:32 pm

UN-Habitat (2011) Cities for all: bridging the urban divide-state of the world's cities 2010/2011 by UN-Habitat. In: First published by Earthscan in the UK and USA in 2008 for and on behalf of the United Nations Human Settlements Programme (UN-HABITAT). http://unhabitat.org/books/state-of-the-worlds-cities-20102011-cities-for-all-bridging-the-urban-divide/

UN Statistical Division (2021) Informal settlement-SDG indicators. https://unstats.un.org/sdgs/report/2021/goal-11/. Accessed 07 Feb 2023 at 12:06 pm

Waliuzzaman SM, Alam A (2022) Commoning the city for survival in urban informal settlements. Asia Pac Viewp 63(1):97–112. https://doi.org/10.1111/apv.12332

Winans K, Kendall A, Deng H (2017) The history and current applications of the circular economy concept. Renew Sust Energ Rev 68:825–833. https://doi.org/10.1016/j.rser.2016.09.123

World Bank (2022) Urban development overview. https://www.worldbank.org/en/topic/urbandevelopment/overview#1. Accessed 09 Feb 2023 at 09:14 am

Isaac Nevis Fianoo holds a Master's degree in Sustainability Science at the United Nations University—Institute for the Advanced Study of Sustainability (UNU-IAS), Tokyo-Japan. Prior to his master's degree, he served as a research and teaching assistant at the University for Development Studies. He is an intern at the Satoyama Initiative International Partnership (IPSI). His research interest lies in land use/spatial planning, local governance, environmental policy and planning, urban resilience, and community participation.

Stephen Leonard Mensah is a Ph.D. candidate in Urban Affairs at the University of Memphis. He is a community-focused interdisciplinary scholar with research interests in urban and community sustainable development from multidimensional perspectives, including climate urbanism, housing, circular economy, residents' everyday experiences, and policy responses to inequities in local environments. He is particularly interested in exploring these themes using the lens of inclusive, resilient, and sustainable development and the application of varying research methods.

Dr. Seth Asare Okyere is an interdisciplinary development planner working at the intersection of international development, sustainable urbanism, community resilience, and social equity. He has extensive experience across academia, policy agencies, and non-governmental organisations conducting transformative research, leading international expert teams, coordinating international joint projects, and engaging local communities to build capacity for sustainable development. His experiences span cities and communities in Africa, Europe, Asia, and North America. Seth is in pursuit of the sustainable development of vulnerable communities within mission-focused approaches to co-producing equitable, resilient, and place-based solutions. Seth holds a Ph.D. in Urban Development Planning (Osaka University, Japan) and an MSc in Urban Planning and Policy Design (Polytechnic University of Milan, Italy). He currently serves as a Visiting Assistant Professor (University of Arizona, USA) and an adjunct Associate Professor (Osaka University, Japan).

Matthew Boadie-Ampong is currently a doctoral student in Sustainability Science at the United Nations University, Tokyo-Japan. He holds two master's degrees in political analysis and Public Policy from the National Research University Higher School of Economics (Moscow, Russia) and Development Policy with a specialization in Sustainable Development from the Korea Development Institute School of Public Policy and Management (Republic of Korea). His research interests include the Politics and Policies of Sustainable Urban Governance, Human well-being, Biodiversity, and Society, Sustainable Business Development and Management, and Education for Sustainability.

Joshua Jirjiri is a field researcher, project coordinator, and certified US Government YALI fellow. Joshua has a B.A. in Integrated Community Development from the University for Development Studies (UDS), and he is currently enrolled at the University of Business and Integrated Development Studies (UBIDS), pursuing an MPhil in Development Management. Joshua's research focuses on water governance and policy. He evaluates the efficiency of existing water governance frameworks, rules, and laws and looks for ways to enhance water governance for just and sustainable water management. To advance water governance, management, and the cause of evidence-based decision-making, Joshua is interested in interdisciplinary collaborations, interaction with policymakers, and participation in international conferences.

Modern Vernacular Architecture and Circular Economy in Informal Settlements

Johan Mottelson

Abstract This chapter investigates circular economy practices in the informal housing construction sector based on case studies situated in Africa, Asia, and South America. The examined cases highlight the application of circular material use within the informal settlements largely based on downcycling, first through reuse and subsequently recycling. In addition, one case exemplifies upcycled waste from the formal city used as interior and exterior cladding of housing. Moreover, the chapter highlights deliberate design strategies for dwelling expansion that enable housing adaptations to accommodate future household needs according to the available economic resources, underscoring distinct approaches to the circular economy in dwelling design that constitute forms of resilient urbanism. The presented cases underscore that the use of reclaimed materials for construction is a notable common practice in informal settlements, particularly among low-income groups, although almost all housing across all cases were made of virgin industrial construction materials. The findings underscore that traditional vernacular architecture is a relatively marginal phenomenon in contemporary urban areas in the Global South, as industrial construction materials such as corrugated iron sheets and reinforced concrete have proliferated based on market mechanisms.

Keywords Vernacular architecture · Modernist architecture · Informal housing · Informal settlements · Circular economy

1 Introduction

Informal settlements are populated areas that operate and develop with limited state control (Dovey and King 2011; Jenkins 2006). In this context, informal housing refers to dwellings constructed without state authorization, compliance with the building

J. Mottelson (✉)
Institute of Architecture, Urbanism and Landscape, Royal Danish Academy, Copenhagen, Denmark
e-mail: jmot@kglakademi.dk

S. A. Okyere et al. (eds.), *Urban Slums and Circular Economy Synergies in the Global South*, Advances in 21st Century Human Settlements,
https://doi.org/10.1007/978-981-99-9025-2_3

code, and formal rights to the land (Dekel 2020). More than one billion people reside in informal settlements, mostly in the Global South, reflecting a failure of the formal housing markets to provide affordable accommodation (Angel 2000; Satterthwaite et al. 2020). Informal housing markets are able to provide accommodation at a lower price than formal markets as the regulatory standards and minimum requirements for construction are not complied with (Visagie and Turok 2020). Informal housing typically proliferates in contexts characterized by high levels of urban population growth, limited state capacity to administer the urban growth, limited economic resources to comply with regulation, and unrealistic regulatory requirements (Lusugga Kironde 2006).

There are a multitude of issues associated with the lack of state recognition, lack of enforcement of building and planning regulations, and economic deprivation typical of informal settlements, including the risk of evictions, inadequate access to basic infrastructure, and sub-standard housing (Satterthwaite et al. 2020). Terminology used to describe informal housing such as 'makeshift structures' or 'shanty towns' highlights the temporary material dimensions of the built environment of informal settlements. While previous research has examined circular economy practices in the informal sector of low-income countries (Korsunova, et al. 2022), limited scholarly works have investigated these processes in the contemporary informal housing construction sector despite the vast scale of this form of accommodation.

The present study investigates construction practices of informal housing in the Global South with an emphasis on circular economy based on three case studies of informal settlements situated in major cities in Africa, Asia, and South America, including Maputo, Mozambique; Hanoi, Vietnam; and Rio de Janeiro, Brazil. The study employs a qualitative research approach informed by the fields of visual anthropology and architectural research (see Collier and Collier 1986; Groat and Wang 2013). This relies on photographic documentation of construction practices with an emphasis on circular economy principles collected through field work in each of the case study areas. The photographic material is analyzed qualitatively and contextualized in the literature on circular economy through which general tendencies across the cases are extrapolated.

The examined cases highlight the application of circular material use within informal settlements, which is largely based on downcycling. These processes typically involve reusing materials and subsequently recycling them. For example, in Maputo, Mozambique, degraded corrugated iron sheet roofs are typically first reused as fencing and subsequently recycled as scrap metal when further degraded. However, a case en Hanoi, Vietnam demonstrates the use of upcycled waste from the formal city as interior and exterior cladding for housing. The chapter highlights deliberate design strategies for dwelling expansion that enable housing adaptations to accommodate future household needs based on available economic resources. These design strategies constitute forms of resilient urbanism and distinct approaches to circular economy in dwelling design.

The presented cases underscore that the use of reclaimed materials for construction is a common practice in informal settlements, particularly among low-income groups. However, the vast majority of housing across all three case study areas were made of virgin industrial construction materials. The findings thus suggest that traditional vernacular architecture is a relatively marginal phenomenon in contemporary urban areas in the Global South, as industrial construction materials such as corrugated iron sheets and reinforced concrete have proliferated based on market mechanisms. On this background, the chapter conceptualizes the emergence of a modern vernacular architecture that is typified by the use of modern industrial construction materials in interplay with local contextual geographic, economic, and cultural conditions with limited involvement of engineers, urban planners, and architects.

2 Circular Economy

A linear economy operates through the extraction of natural resources, manufacturing of products, and ultimately disposal of waste, also known as take-make-dispose or cradle-to-grave (Braungart and McDonough 2009; Goyal et al. 2018). In contrast, a circular economy operates through a closed-loop system by employing reduce, reuse, and recycle or cradle-to-cradle paradigms, through which waste streams are used as inputs in the production to minimize waste generation, as well as reducing pollution and carbon emissions (Braungart and McDonough 2009; Goyal et al. 2018). Circular economy practices in the construction sector have gained increased academic and wider societal attention because of the potentials in addressing the challenges of resource depletion and climate change (Schroeder et al. 2019). However, this requires wide-ranging transformations in the contemporary approaches to building design and construction from linear waste production toward regenerative, closed-loop cyclical material flows. Notable circular economy practices in the construction sector include reduction of waste through optimized material use, using construction materials with low carbon footprints, design for disassembly and reuse, application of adaptative design strategies, and material recovery of waste streams utilized in construction (Joensuu et al. 2020). Many of these approaches were unified through the cradle-to-cradle paradigm which cover design strategies emphasizing the separation of different materials and design for disassembly to enhance the potentials for reusing and recycling the materials (Braungart and McDonough 2009).

Waste valorization is a critical strategy to enhance circular material flows, whereby waste materials become inputs for other processes through economic incentives (Ferronato et al. 2019). Upcycling and downcycling are two important approaches to waste valorization and reuse or recycling of materials. Upcycling refers to the process of transforming waste materials into new products of higher value or quality (Singh et al. 2019). An example of upcycling is the use of textile waste in the production of fashionable products with a high retail value (Cassidy and Han 2013). Downcycling refers to the process of reusing or recycling waste materials into products of lower value or quality (Joensuu et al. 2020). This process often involves breaking down materials into their component parts, which are then used to create new products that may not be as durable or valuable as the original material. An example of downcycling is the reuse of textile fibers in thermal insulation of walls (Sandin and Peters 2018).

While much of the scholarly literature has emphasized the potentials of circular economy practices in high-income countries to address unsustainable resource consumption dynamics, a growing body of literature has emphasized circular economy practices in low-income countries (Ferronato et al. 2019). Notably, Korsunova et al. (2022) introduce the concept of necessity-driven circular economy, describing a set of formal and informal practices aimed at restoring and retaining the value of goods and materials based on economic necessity and income generation. However, research on circular economy practices in the construction sector of low-income countries remains limited.

3 Vernacular and Modernist Architecture

Vernacular architecture refers to traditional or indigenous building styles and techniques that have developed over time in response to local environmental, cultural, and social conditions (Oliver 2007b). Vernacular architecture is typified by little to no involvement of the formal architecture profession and it emerges through bottom-up and trial-and-error processes (Rudofsky 1987). Vernacular architecture largely employs locally available materials, such as mud, timber, or stone, and utilizes designs that reflect the unique characteristics of the surrounding landscape and climate as well as the local social and cultural context (Fathy 1986). Vernacular architecture can be found all over the world, from the adobe houses of West Africa to the thatched roof cottages of rural England (Oliver 2007b). These structures often incorporate sustainable design principles, such as natural ventilation and passive solar heating, which are well-suited to the local climate and help to reduce energy consumption and environmental impact (Fathy 1986). In addition to their environmental benefits, vernacular buildings also reflect the cultural identity and history of the communities that build them. Many vernacular buildings incorporate traditional patterns, materials, and design elements that have been passed down through generations, creating a sense of continuity and connection to the past (Glassie 2000). Despite the many advantages of vernacular architecture, it is often threatened by the spread of modern construction techniques and the global homogenization of architectural styles (Oliver 2007a). This development has occurred in part because of lower costs and reduced need for maintenance, as well as cultural preferences that often favor the use of industrial construction materials. However, there is a growing interest in preserving and promoting vernacular architecture as a means of promoting sustainability, since vernacular architecture is typically based on locally sourced materials with low carbon footprints.

Modernist architecture emerged in the late nineteenth and early twentieth century in response to the social, economic, and technological changes brought about by industrialization (Benevolo 1977). The modernist movement was largely driven by formally trained architects, engineers, and industrialists through top-down principles. Modernist architects largely rejected the historical styles of the past and instead focused on simplicity, functionality, and the use of new industrial materials and construction techniques (Curtis 1996). This led to the development of the International Style, a movement characterized by rectilinear forms, glass facades, as well as concrete and steel structures applied through largely universal principles irrespective of context (Hitchcock and Johnson 1995). The style was pioneered by high-profile architects in the first half of the twentieth century, such as Le Corbusier, Walter Gropius, and Mies van der Rohe, who believed that modern architecture had a moral and social responsibility to improve the lives of people by creating efficient, healthy, and beautiful living spaces (Benevolo 1977). Despite the many innovations associated with modernist architecture, the International Style has been criticized for its lack of contextual sensitivity and its imposition of a standardized aesthetic on different cultures and environments as well as homogenization of the global built environment, erasing local traditions and vernacular styles (Oliver 2007a). Accordingly, the modernism movement arguably represents an antithesis to vernacular architecture with its grand schemes for urban development imposed through top-down processes and highly industrialized large-scale approach to construction. More recently, the high level of energy consumption associated with industrial construction materials has become a subject of debate and a controversial aspect of modernist approaches to design and construction due to the global climate crisis and the considerable greenhouse gas emissions associated with the construction sector. However, the construction techniques and material use pioneered by the modernist movement have largely been universally adopted worldwide and continue to predominate across contemporary construction sectors, thus accounting for the vast majority of the volume of construction materials used and carbon emissions associated with the construction sector.

4 Case Studies

This chapter examines three case studies of informal settlements situated in different parts of the Global South, including Khâm Thiên, Hanoi, Vietnam; Maxaquene, Maputo, Mozambique; Rocinha, Rio de Janeiro, Brazil. The locations of each of the cities are presented in Fig. 1. The three settlements were selected as case studies for the paper as they are located on different continents across the Global South and thereby provide a broad international scope on conditions of informal settlements across highly heterogeneous contexts. Secondly, all three areas are centrally located in major cities in each respective context, and thereby represent some of the most densely populated and congested settlements, making them appropriate cases for exploring the boundaries and challenges imposed by their specific contextual conditions. Despite the contested legality associated with these settlements, most households across all three cases have access to water, sanitation, and electricity. Furthermore, sub-standard housing and overcrowding are limited in these cases.

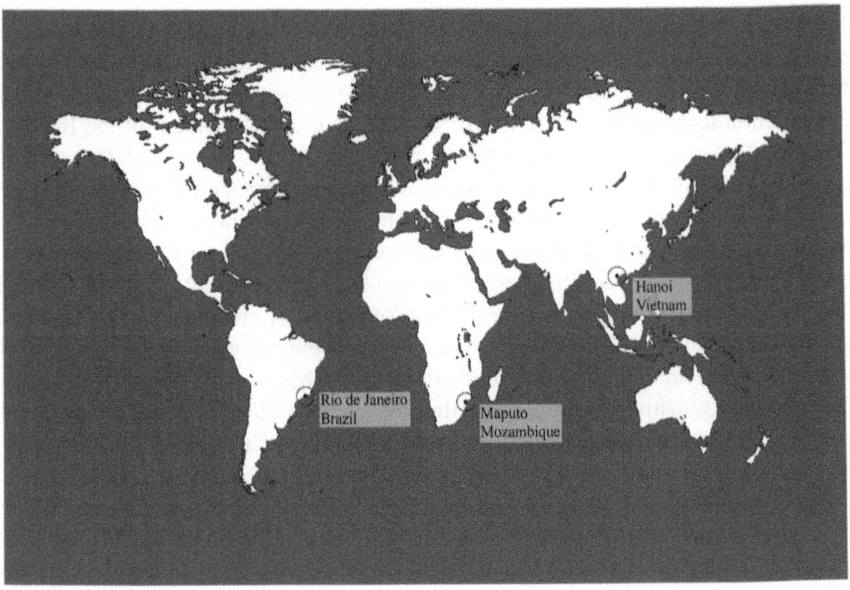

Fig. 1 Location of the cities in which the case studies are situated

Accordingly, so-called 'slum'-characteristics are limited (see UN-Habitat 2003) and the settlements accommodate both middle-class residents and low-income groups. This social diversity and moderate levels of deprivation are not unusual in informal settlements in the respective contexts and the cases are thus relatively representative of centrally located settlements in each setting. Nevertheless, since landownership is contested and building and planning regulations are enforced to a limited degree across the three contexts, all the case study areas are informal settlements by definition. Photos of the urban fabric as well as typical access pathways of each settlement examined in this chapter, are presented in Figs. 2 and 3.

4.1 Khâm Thiên, Hanoi, Vietnam

Hanoi is the capital of Vietnam and the political, economic, and cultural center of the country (Van Horen 2005). The city is located in the delta of the Sông H´ông river and has around 7 million inhabitants (Leducq and Scarwell 2018). In the aftermath of the Vietnam war and the political instability of the country in the late 1970s, many of the major informal settlements in Hanoi were established (Logan 2000). Today, much of the contemporary urban development in the city is regulated by authorities (Geertman and Kim 2019; Van Horen 2005). Khâm Thiên is a centrally located, large informal settlement in Hanoi (see Fig. 4). The neighborhood is one of the most densely built-up urban fabrics in the city. While some of the neighborhood was developed as a

Fig. 2 An overview of the built environment of **a** Khâm Thiên (Hanoi, Vietnam), **b** Maxaquene (Maputo, Mozambique), and **c** Rocinha (Rio de Janeiro, Brazil)

social housing scheme in the 1970s, much of it developed subsequently without urban planning and enforcement of regulation (see Hong and Kim 2021). Informal housing in Khâm Thiên is typically 2–4 m wide with a toilet, kitchen, and stairs in the back of the house and a single façade facing the street. The typical housing typology may have been inspired by traditional narrow housing in the historic city center known as 'tunnel' or 'tube' houses (see Kien 2008).

Fig. 3 Typical street spaces in Khâm Thiên (left), Maxaquene (middle), and Rocinha (right)

Fig. 4 Location of Khâm Thiên in Hanoi, Vietnam

4.2 *Maxaquene, Maputo, Mozambique*

Located on the coast of the Indian Ocean, Maputo is the capital of Mozambique and the political, economic, and cultural center of the country. The urban development after independence in 1975 was characterized by a large influx of rural population (Jenkins 2000). The ongoing civil war resulted in a large number of internally displaced people and the abolishment of previous restrictions on the movement of

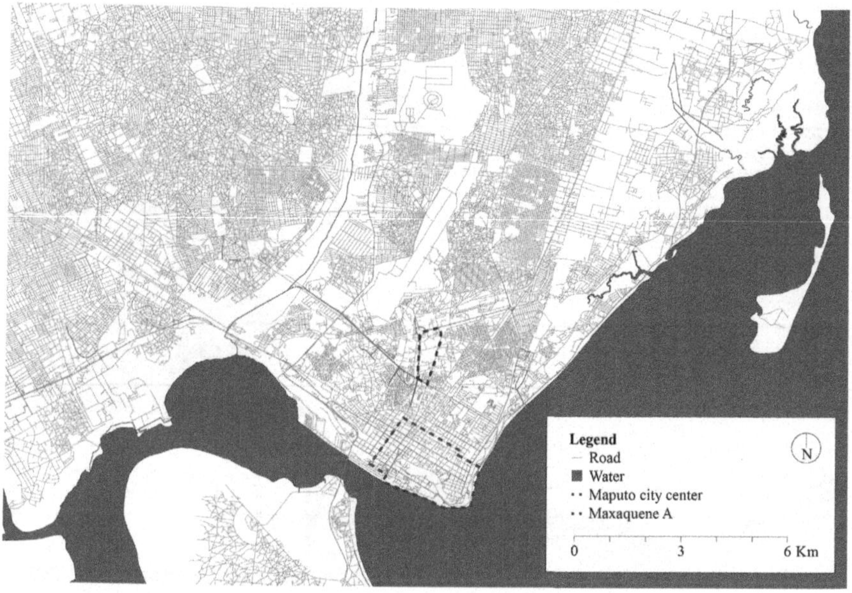

Fig. 5 Location of Maxaquene in Maputo, Mozambique

people led to rapid urbanization. During this period many of the centrally located informal settlements were established (Jenkins and Mottelson 2020). The informal urban development continues today, as the urban growth largely occurs without urban planning and enforcement of regulation (Mottelson 2020). Before Independence, Mozambicans were generally only allowed to construct houses of impermanent building materials. Today, most housing is built of concrete. Maxaquene is one of the most centrally located informal settlements in Maputo (see Fig. 5). The settlement was largely established in the 1970s and 1980s (Saevfors 1986). The majority of the neighborhood consists of single-story detached housing with private outdoor areas. The incremental development of dwellings with single-room horizontal expansions built according to optimized use of the standardized construction materials has led to the emergence of the 'Casa Ventoinha' (fan house) housing typology in southern Mozambique (Carrilho 2004). The typology resembles a fan when seen from above with four individual roof surfaces, expressing the interior division of the rooms.

4.3 Rocinha, Rio de Janeiro, Brazil

Rio de Janeiro is the second-largest city in Brazil and one of the most populous cities in Latin America (Snyder et al. 2014). Most informal settlements (known as favelas in Brazil) in Rio de Janeiro emerged in the 1960s and 1970s during a period of rapid urbanization. During this time, authorities attempted to evict several

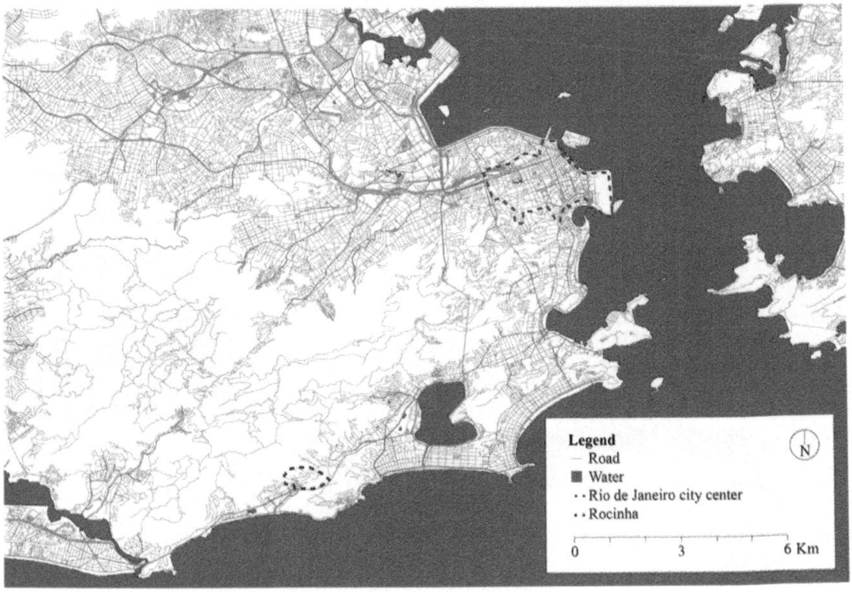

Fig. 6 Location of Rocinha in Rio de Janeiro, Brazil

parts of centrally located favelas and relocate residents to peripheral areas of the city (Fahlberg et al. 2020). When Brazil transitioned to democracy in the 1980s, some property rights were granted to the residents of favelas and the government began to invest in basic infrastructure in these informal settlements (Friendly 2017). Today, the authorities largely accept the informal urban development within the established favelas while urban development outside these areas largely occurs within the formal legal framework (Caldeira and Holston 2016). Rocinha is the largest informal settlement in Brazil, located on a steep hill facing the ocean (see Fig. 6) (Freire-Medeiros 2009). The early formation of the settlement began in 1920s and 1930s and was characterized by simple makeshift housing (Abreu 1987). Today, the settlement is densely built up with multi-story concrete structures.

5 Methods

This study employed a qualitative research design to investigate and document circular economic practices in the construction of informal housing in three distinct regions. The research design was informed by the fields of visual anthropology and architectural research (see Collier and Collier 1986; Groat and Wang 2013), enabling an exploration of the built environment through photographic documentation which served as the primary method of data collection. The photographs were used to capture the physical manifestations of circular economy principles within informal housing structures observed in each neighborhood, providing visual evidence of material reuse, adaptive construction techniques, and resource optimization. To ensure a comprehensive coverage of the built environment, multiple site visits were conducted to each case study location. The photographic documentation was curated to capture different stages of construction, architectural elements, and adoption of circular economy practices. The visual analysis involved a systematic categorization of the images based on material reuse, local craftsmanship, and innovative construction techniques. This process highlighted key features, generalizable patterns, and particular examples of circular economy practices within the built environment across the different settings. This enabled the identification of the underlying principles of circular economy in the construction of informal housing across the three contexts while highlighting intercontinental macro-level dynamics characterizing the examined construction sectors.

6 Field Observations on Circular Economy

Although all three case study settlements are situated in different parts of the world, the built environments across the three contexts are characterized by a high degree of commonalities. Almost all the structural elements of the entire building stock of all three settlements are comprised of concrete (columns, beams, and slabs) with infill walls of concrete blocks or bricks (see Figs. 2, 3 and 7). The building roofs are comprised almost exclusively of corrugated iron sheets or concrete slabs. Photos of typical construction sites in each case study area are presented in Fig. 7.

In Rocinha and Khâm Thiên, the top floor of each housing is usually left without complete walls and used for water tanks, laundry, and recreational purposes. Rebars of the concrete structures are often exposed on the top floors, enabling potential

Fig. 7 Construction sites in Khâm Thiên (left), Maxaquene (middle), and Rocinha (right)

Fig. 8 Adaptable design strategies in Rocinha

vertical expansions of the dwellings. The top floors typically feature an open steel structure which can be de- and remounted on top of the building should the family seek to expand the dwelling vertically through the addition of another concrete floor. Photos of these practices are presented in Figs. 2 and 8.

In Maxaquene, the built environment is almost exclusively single-story. In this context, verticalization of housing is generally more costly than horizontal expansion due to the increased structural requirements (Mottelson and Andersen 2018). As most plots have open outdoor spaces suitable for horizontal expansion of the dwellings, this has become the common practice and is closely linked to the common housing typology, *casa ventoinha,* which is characterized by this single-room incremental horizontal expansion of the dwelling. This practice is observable in Fig. 2. All three cases thus highlight distinct design strategies that enable incremental expansion of the dwellings corresponding to the household size and economy. Such strategies exemplify circular economy approaches to housing design to enhance the adaptability of the housing stock to accommodate future needs.

In Maputo, most housing is single-story, and roofs are typically made of timber and corrugated iron sheets. A notable feature is the reuse of old corrugated iron sheet roofs in the fences used for delineating plots. Lines of holes are observable in these, indicating previous joinery to timber rafters. These fences are often allowed to

Fig. 9 **a** Makeshift structure partially made of reclaimed asbestos cement boards, **b** Collection of used plastics for recycling

degrade over time and ultimately become scrap material which is routinely collected in steel carts (tchovas) and transported to a steelyard, where it is sold by weight for recycling. This is an example of downcycling, first through repurposing of a material component in a function of lower value and ultimately recycling, whereby the scrap metal is used for the production of new products. A photo of corrugated iron roof sheets repurposed as fencing is presented in Fig. 10. Similarly, plastic waste is widely collected, sorted, and transported to a major plastic factory in the city which purchases the plastic by weight (see Fig. 9). The sorting involves separating bottles (PET), plastic bags (LDPE), washbasins (HDPE), containers, and furniture (PP) to enable recycling and differentiate valorization of different plastic types. Notably, several instances of the use of reclaimed asbestos cement boards were observed in the neighborhood, potentially posing health hazards (see Fig. 9).

A distinctive part of Khâm Thiên is situated on the swampy shores of a local lake in which wastewater is discharged and garbage is routinely dumped. Here, the dwellings are mostly single-story and largely characterized by rental housing for low-income groups.[1] Most of these dwellings feature extensive use of repurposed materials (see Fig. 12). In one instance, the façade of an entire single-room dwelling

[1] Disclosed through interviews with local residents.

Fig. 10 Downcycling of construction materials in Maxaquene (left) and Khâm Thiên (right)

is made up of old doors with traces of previous use (see Fig. 10). Doors are usually more expensive than façade cladding materials, and as such, this represents a case of downcycling. Several of the neighboring dwellings feature extensive use of different plywood, particle board, timber, and composite materials (see Fig. 12). These may be considered upcycling of waste materials. A notable and innovative use of repurposed materials is found inside several of the dwellings, namely, the use of old advertisement billboards for interior wall cladding (see Fig. 11). These are generally made of hollow polypropylene plastic boards or heavy vinyl tarp, providing durable, flexible, waterproof materials for wall cladding. This is a notable example of upcycling of waste materials used for a higher value purpose.

7 Discussion

The field observations highlight an notable examples of different circular economy practices in the informal housing construction sectors across the examined cases in Africa, Asia, and South America. The dwellings in all three contexts feature distinct design strategies that enable housing expansion over time and thereby enhance the adaptability of the dwellings to accommodate future needs. In Khâm Thiên and Rocinha, this is enabled through exposed steel reinforcement in the concrete structures and flexible steel structures on the top floor of the dwellings which permit vertical expansion of the dwellings (see Figs. 2 and 8). In Maxaquene, dwellings are expanded horizontally one room incrementally (see Fig. 2). These strategies represent circular economy approaches to dwelling design, which enhances the adaptability of the housing and thus constitute highly resilient urban development approaches in response to the local economic and spatial conditions, through which housing can be expanded gradually when the household economic conditions permit it.

The field observations furthermore exemplify different modes of downcycling of construction materials (i.e., repurposing or recycling of a material component for a function of lower value) (see Schroeder et al. 2019). In Maputo this is exemplified through the reuse of old corrugated iron sheet roofs for fencing, and once the metal sheets are further degraded, these are eventually collected and sold by weight for recycling, similarly to the plastic waste collection seen in much of the city (see Fig. 10). Similarly, repurposed materials such as old doors used for façade cladding in Hanoi demonstrate downcycling practices in the construction sector of informal settlements (see Figs. 10 and 12).

Notably, the field observations also highlight innovative upcycling practices (see Singh et al. 2019). These were only found in Hanoi, where scrap materials and advertisement billboards were repurposed as interior and exterior wall cladding (see Fig. 11). These underscore valorization of waste materials streams that largely emanate from the formal city. The examined examples of up- and downcycling suggest that within informal settlements, construction materials are likely largely downcycled, whereas examples of upcycling of waste materials generally stem from the formal city. Nevertheless, the findings also highlight the use of asbestos cement boards (see Figs. 9 and 12). The use of asbestos cement boards in the formal construction sector has been widely abolished due to restrictive regulations adopted as inhalation of asbestos fibers can lead to life-threatening lung conditions, such as mesothelioma, asbestosis, and lung cancer. These observations thus highlight some of the negative externalities associated with circular economic practices in an unregulated context.

The field observations underscore that although circular economy approaches are widespread, circular material usage in the built remain limited considering the general scope of the built environment. Most construction in the three settlements is based on virgin materials. Structurally, virtually the entire building stock of all three examined settlements were based on reinforced concrete structures along with some use of steel. The exemplified cases of reuse of old construction materials all comprise cases of light weight sheets used for façade or interior cladding of relatively limited volume and weight. Notably, few examples of repurposed materials were used as structural components of construction, underscoring the somewhat limited application. In addition, circular material use was almost exclusively observed among low-income residents within the settlements, suggesting that it is associated with limited household economic conditions. These practices thus underscore necessity-driven circular economy practices through retainment of the value of scrap materials as a mode of income generation (see Korsunova et al. 2022).

The general field observations across the three examined settlements underscore that traditional vernacular architecture is a relatively marginal phenomenon in contemporary urban areas in the Global South, as industrial construction materials

Fig. 11 Upcycling of waste materials used for interior wall cladding in Khâm Thiên

such as corrugated iron sheets, steel, and reinforced concrete were applied extensively across all three contexts. However, the observations suggest that this architecture is highly responsive to local conditions, as circular economy design strategies have evolved in all three contexts to enable incremental expansions of dwellings according to household economic conditions and family size. Although industrial construction materials make up the majority of the built environment in contemporary informal settlements, the design of dwellings is optimized according to the local geographic, economic, and cultural conditions with limited involvement of academically trained engineers, urban planners, and architects. Accordingly, this bottom-up approach to dwelling design and urban development share distinct characteristics with traditional vernacular architecture. Namely, sophisticated adaptation to local contextual geographic, economic, and cultural conditions. As this emerging trend in contemporary informal settlements utilizes modern industrial construction materials and techniques applied through similar processes as vernacular architecture, this trend arguably epitomizes a *modern vernacular architecture*.

As the built environment in informal settlements is largely unregulated, it remains questionable to what extent building and planning regulations hold the potential to address some of the externalities associated with circular economic practices or to support the wider adoption of circular economy practices. In both Maxaquene and Khâm Thiên, reclaimed asbestos cement boards were observed as notable material components in makeshift structures, posing considerable health risks for the people involved in constructing them as well as the end users. As the institutional capacity and political will to regulate informal settlements is limited across much of the

Fig. 12 **a** Rental housing area in Khâm Thiên largely built of reclaimed materials, **b** Façade made of reclaimed materials, **c** Asbestos cement boards stored next to dwelling

Global South, the building code and planning regulations likely hold limited potential to restrict the use of toxic or otherwise hazardous materials or practices. Instead, adapting current incentive structures to support sustainable material use likely holds more potential. Such measures may include subsidies for the collection and recycling of materials (e.g., scrap steel components and used plastics) to further promote such current common practices in the analyzed settlements operating on a market basis. Furthermore, providing a state-subsidized refund on toxic materials will likely enable the collection of these through bottom-up measures, similar to the current collection of used plastics and scrap steel components. In all three analyzed settlements, biodegradable and inorganic waste streams were not separated. Separation of organic and inorganic matter may hold potential for fuel production utilizing the organic waste streams. In terms of measures for advancing the use of recycled or reclaimed materials for construction, further analysis is required to identify potential valorization of waste streams.

8 Conclusion

Although over one billion people worldwide is accommodated in informal housing typified by forms of contested legality, limited research has examined the building material characteristics of this major global trend. This chapter investigates the material conditions of informal housing, with an emphasis on circular economy principles. The study draws on case studies from Hanoi, Vietnam; Maputo, Mozambique; and Rio de Janeiro, Brazil, to illustrate the application of circular material use and circular economy approaches to dwelling design in the context of informal settlements. The presented cases underscore that the use of reclaimed materials for construction is a notable common practice across all three examined cases despite their diverse geographic settings in Africa, Asia, and South America. These settlements feature distinct design strategies that allow for the expansion of dwellings over time, enhancing adaptability to future needs. The circular economy approaches to dwelling design enable gradual expansion of housing when household economic conditions permit it and thus underscore a notably resilient urban development tendency.

The chapter also highlights that circular material flows within informal settlements are largely based on downcycling, first through reuse and subsequently recycling. This is exemplified through the use of old corrugated iron roof sheets as fencing and subsequently recycling. Notably, innovative upcycling practices were found in Hanoi, where scrap materials and advertisement billboards were repurposed for interior and exterior wall cladding. However, despite the widespread use of circular economy approaches, circular material usage in the built environment remains limited. Most construction in the settlements is based on virgin industrial construction materials, and the reuse of old construction materials is mainly limited to lightweight sheets used for façade or interior cladding of relatively limited volume and weight.

The chapter suggests that traditional vernacular architecture, which relies on locally sourced and sustainable materials, is relatively marginal in contemporary urban areas in the Global South. Instead, industrial construction materials such as corrugated iron sheets, steel, and reinforced concrete have become dominant based on market mechanisms. On this background, the chapter conceptualizes the emergence of a new type of vernacular architecture in contemporary informal settlements. This architecture is both modern and vernacular, as it is based on modern industrial construction materials applied in interplay with local contextual geographic, economic, and cultural conditions, with limited involvement of academically trained engineers, urban planners, and architects. The chapter labels this modern vernacular architecture as it features industrial material use characteristic of modern architecture applied in interplay with local conditions through bottom-up processes similar to vernacular architecture.

References

Abreu MDA (1987) Evolução urbana do Rio de Janeiro

Angel S (2000) Housing policy matters: a global analysis. Oxford University Press

Benevolo L (1977) History of modern architecture, vol 2. Mit Press

Braungart M, McDonough W (2009) Cradle to cradle. Random House

Caldeira T, Holston J (2016) State and urban space in Brazil: from modernist planning to democratic interventions. In: Managing urban futures. Routledge, pp 159–180

Carrilho J (2004) Traditional informal settlements in Mozambique: from Lichinga to Maputo. FAPF

Cassidy TD, Han SL-C (2013) Upcycling fashion for mass production. In: Sustainability in fashion and textiles. Routledge

Collier J, Collier M (1986) Visual anthropology: photography as a research method. UNM Press

Curtis WJ (1996) Modern architecture since 1900. Phaidon London

Dekel T (2020) The institutional perspective on informal housing. Habitat Int 106:102287. https://doi.org/10.1016/j.habitatint.2020.102287

Dovey K, King R (2011) Forms of informality: morphology and visibility of informal settlements. Built Environ 37(1):11–29. https://doi.org/10.2148/benv.37.1.11

Fahlberg A, Vicino TJ, Fernandes R, Potiguara V (2020) Confronting chronic shocks: social resilience in Rio de Janeiro's poor neighbourhoods. Cities 99:102623. https://doi.org/10.1016/j.cities.2020.102623

Fathy H (1986) Natural energy and vernacular architecture

Ferronato N, Rada EC, Gorritty Portillo MA, Cioca LI, Ragazzi M, Torretta V (2019) Introduction of the circular economy within developing regions: a comparative analysis of advantages and opportunities for waste valorization. J Environ Manage 230:366–378. https://doi.org/10.1016/j.jenvman.2018.09.095

Freire-Medeiros B (2009) The favela and its touristic transits. Geoforum 40(4):580–588

Friendly A (2017) Urban policy, social movements, and the right to the city in Brazil. Lat Am Perspect 44(2):132–148

Geertman S, Kim B (2019) Hanoi: a study of informally developed housing and its role in the political arena of a post-reform communist city. In: The Routledge handbook on informal urbanization. Routledge, pp 112–123

Glassie H (2000) Vernacular architecture. Indiana University Press

Goyal S, Esposito M, Kapoor A (2018) Circular economy business models in developing economies: lessons from India on reduce, recycle, and reuse paradigms. Thunderbird Int Bus Rev 60(5):729–740. https://doi.org/10.1002/tie.21883

Groat LN, Wang D (2013) Architectural research methods. Wiley, pp 215–263

Hitchcock HR, Johnson P (1995). The international style. WW Norton & Company

Hong N, Kim S (2021) Persistence of the socialist collective housing areas (KTTs): The evolution and contemporary transformation of mass housing in Hanoi, Vietnam. J Hous Built Environ 36(2):601–625. https://doi.org/10.1007/s10901-020-09765-1

Jenkins P (2000) City profile: Maputo. Cities 17(3):207–218. https://doi.org/10.1016/S0264-2751(00)00002-0

Jenkins P (2006) Informal settlements: infernal and eternal? The role of research in policy advocacy and urban informal settlements in Angola. In: Huchzermeyer M, Karam A (eds) Informal settlements—a perpetual challenge? University of Cape Town Press, pp 84–102

Jenkins P, Mottelson J (2020) Understanding density in unplanned and unregulated settlements of peri-urban Africa: a case study of Maputo, Mozambique. In: Todes MRA, Harrison P, Appelbaum A (eds) Densifying the city? Global cases and Johannesburg. Edgar Elgar Publishing, pp 116–128. https://doi.org/10.4337/9781789904949.00016
Joensuu T, Edelman H, Saari A (2020) Circular economy practices in the built environment. J Clean Prod 276:124215. https://doi.org/10.1016/j.jclepro.2020.124215
Kien T (2008) "Tube House" and "Neo Tube House" in Hanoi: a comparative study on identity and typology. J Asian Arch Build Eng 7(2):255–262
Korsunova A, Halme M, Kourula A, Levänen J, Lima-Toivanen M (2022) Necessity-driven circular economy in low-income contexts: how informal sector practices retain value for circularity. Glob Environ Chang 76:102573. https://doi.org/10.1016/j.gloenvcha.2022.102573
Leducq D, Scarwell H-J (2018) The new Hanoi: opportunities and challenges for future urban development. Cities 72:70–81
Logan WS (2000) Hanoi: biography of a city. UNSW Press
Lusugga Kironde JM (2006) The regulatory framework, unplanned development and urban poverty: findings from Dar es Salaam, Tanzania. Land Use Policy 23(4):460–472. https://doi.org/10.1016/j.landusepol.2005.07.004
Mottelson J (2020) A new hypothesis on informal land supply, livelihood, and urban form in Sub-Saharan African cities. Land 9(11):11. https://doi.org/10.3390/land9110435
Mottelson J, Andersen JE (2018) Casas melhoradas: affordable housing in Maputo, Mozambique
Oliver P (2007a) Built to meet needs: cultural issues in vernacular architecture. Routledge
Oliver P (2007b) Dwellings: the vernacular house world wide. Phaidon Press
Rudofsky B (1987) Architecture without architects: a short introduction to non-pedigreed architecture. UNM Press
Saevfors I (1986) Maxaquene: a comprehensive account of the first urban upgrading experience in new Mozambique. UNESCO
Sandin G, Peters GM (2018) Environmental impact of textile reuse and recycling–a review. J Clean Prod 184:353–365
Satterthwaite D, Archer D, Colenbrander S, Dodman D, Hardoy J, Mitlin D, Patel S (2020) Building resilience to climate change in informal settlements. One Earth 2(2):143–156. https://doi.org/10.1016/j.oneear.2020.02.002
Schroeder P, Anggraeni K, Weber U (2019) The relevance of circular economy practices to the sustainable development goals. J Ind Ecol 23(1):77–95. https://doi.org/10.1111/jiec.12732
Singh J, Sung K, Cooper T, West K, Mont O (2019) Challenges and opportunities for scaling up upcycling businesses—the case of textile and wood upcycling businesses in the UK. Resour Conserv Recycl 150:104439. https://doi.org/10.1016/j.resconrec.2019.104439
Snyder RE, Jaimes G, Riley LW, Faerstein E, Corburn J (2014) A comparison of social and spatial determinants of health between formal and informal settlements in a large metropolitan setting in Brazil. J Urban Health 91(3):432–445. https://doi.org/10.1007/s11524-013-9848-1
UN-Habitat. (2003). The challenge of slums: global report on human settlements, 2003. United Nations Human Settlements Programme, Earthscan Publications
Van Horen B (2005) Hanoi. Cities 22(2):161–173
Visagie J, Turok I (2020) Getting urban density to work in informal settlements in Africa. Environ Urban 32(2):351–370. https://doi.org/10.1177/0956247820907808

Johan Mottelson, Ph.D. is assistant professor of architecture at the Royal Danish Academy. His work centers on contemporary urbanization dynamics in Eastern Africa and informal urban development in a global context. His research addresses topics such as the impact of spatial planning in contexts with limited institutional capacity to enforce regulation, the interrelation between livelihood and informal housing supply restrictions, and urban densification trends in unplanned settlements. In parallel with his academic work, Mottelson provides design, architecture, and spatial planning consultancy services to private and third sector organizations operating in Eastern Africa.

Exploring Circular Economy Awareness, Perceptions and Practices in Selected Urban Slums in Kigali City of Rwanda

Vanessa Robertson, Eucabeth Majiwa, and Nathan Kanuma Taremwa

Abstract The circular economy (CE) has garnered much attention due to its potential to contribute to sustainable development. While CE initiatives, such as container-based toilets, waste upcycling, biomass energy generation, and circular building materials, hold promise for improving infrastructure and economic opportunities, particularly in slums, there are different and at times contradictory perceptions of the concept of CE. Leveraging CE potentials thus requires a better understanding of CE awareness and perceptions. To explore CE in the context of Rwandan slums, this book chapter sought to answer the following two research questions: *What is the level of CE awareness and which perceptions prevail in Kigali's slums? What are the key CE-related practices in Kigali's slums?* Fifteen key informant interviews and 12 focus group discussions were conducted with stakeholders in the slums of Kigali's districts Kicukiro, Gasabo, and Nyarugenge. The findings reveal that circularity is present in the slums, albeit informally, with little emphasis on business and entrepreneurship. Repurposing emerged as the most prominent circular practice. Besides, the study uncovered a significant skills gap in waste sorting, which is critical as Rwanda seeks to implement a national policy requiring waste sorting at the source.

Keywords Rwanda · Slums · Informal settlements · Attitudes · Awareness · Waste sorting · Circular economy

V. Robertson
Neu-Ulm University of Applied Sciences (HNU), Neu-Ulm, Germany
e-mail: Vanessa.Robertson@hnu.de

E. Majiwa
Jomo Kenyatta University of Agriculture and Technology (JKUAT), Nairobi, Kenya
e-mail: eucamajiwa@rpe.jkuat.ac.ke

N. K. Taremwa (✉)
University of Rwanda (UR), Kigali, Rwanda
e-mail: nk.taremwa@gmail.com

S. A. Okyere et al. (eds.), *Urban Slums and Circular Economy Synergies in the Global South*, Advances in 21st Century Human Settlements,
https://doi.org/10.1007/978-981-99-9025-2_4

1 Introduction

Slums are densely populated neighbourhoods in metropolitan areas and are often faced by challenges such as lack of urban planning, weak socioeconomic status, and an informal economy (D'Aquino et al. 2022; Ukpanyang et al. 2022). In developing countries around the globe, over 1.5 billion people live in urban slums under precarious conditions and by 2030 that number is projected to significantly increase to three billion (UN-Habitat 2020). With more than half of its population living in slums, sub-Saharan Africa (SSA) has the highest percentage of slum dwellers globally (UN-Habitat 2020). Circular economy (CE) concepts have been proposed to improve slum economies. CE refers to a production and consumption model that involves sharing, reusing, repairing, and recycling current materials and products to increase their lifespan (Sijtsema et al. 2020). This model has been advocated as a solution for achieving environmental, social, and economic sustainability by fostering an interdependent and regenerative economic system (Esken et al. 2018).

CE can benefit slum areas due to its potential to induce sustainable economic development and alleviate waste management challenges. Case studies of CE initiatives in informal settlements include biomass-derived energy generation (D'Aquino et al. 2022), container-based toiled solutions (Couder and Kibuthu 2020), and circular building materials (Dey and Iulo 2021). Further, well-managed circular policies such as extended producer responsibility or the polluter pay principle can lead to affordable waste management and hygiene (Allam and Jones 2018). However, existing practices frequently involve polluting and health-compromising actions (ibid). On the positive side, labour-intensive circular practices such as collecting, treating, and remanufacturing have the potential to generate employment (Allam and Jones 2018). Circularity is not a new concept in the Global South, but it has been widely practiced in pre-industrial eras, and circular routines are still widespread in SSA (Schröder et al. 2019). Nevertheless, the Global South remains largely underrepresented in CE research.

Policymakers and practitioners place high hopes in CE, while it is not clearly defined, difficult to measure, and its conceptions vary (Gregson et al. 2015). Conflicting meanings and exaggerated expectations prevail (Lehmann et al. 2023). Most of the existing CE research focuses on operational and technical aspects of physical material and energy flows in production-consumption systems (Korhonen et al. 2018; Corvellec et al. 2022). Existing CE literature primarily addresses practical metrics, instruments, and techniques, while paying less attention to its fundamental assumptions regarding values, societal structures, cultures, worldviews, and paradigmatic potential. However, production and consumption models are intimately connected with societal values and cultural norms (Pál 2022), so that a CE research stream concerned with consumer behaviour and socio-cultural aspects is emerging (van Langen et al. 2021). So far, the perspectives of marginalized groups in the Global South have received little attention in this research stream.

Previous studies on consumer perspectives of the CE have primarily examined specific solutions and assessed consumer acceptance of particular product types or functions (Camacho-Otero et al. 2018). Notably, the studies often concentrate on sectors like fashion and electronics, such as investigating consumer acceptance of refurbished mobile phones (Van Weelden et al. 2016; Vehmas et al. 2018). These studies have identified factors influencing consumer acceptance, such as lack of awareness, misconceptions about refurbishment processes, and perceived trade-offs between risks and benefits. Overall, these studies indicate that consumer acceptance of the CE is influenced by personal characteristics (e.g., materialism), product and service offerings (e.g., quality), knowledge and understanding, experiences, social aspects (e.g., privacy concerns), perceptions of risks and uncertainties, benefits, and other psychological factors (e.g., social norms) (Camacho-Otero et al. 2018).

Little is known about the role of CE in slums and how it is perceived by inhabitants. Therefore, this chapter investigates CE awareness and perceptions as well as existing circular practices in underprivileged neighbourhoods. The present study's goal is to deepen our understanding of CE consumer perceptions and circular practices in slums. We investigate how consumers view the CE at both an abstract and more specific level using 15 key informant interviews and 12 focus group discussions with stakeholders in the three districts of Kigali, the capital city of Rwanda, namely slum areas in the city's districts Kicukiro, Gasabo, and Nyarugenge.

Rwandan slums are particularly interesting for CE research since Rwanda is among the most advanced countries in terms of CE awareness, cleanliness, and environmentally focused policies (Ogutu et al. 2023). Even though it belongs to the least developed countries, Rwandans are well-known for taking a highly proactive stance towards environmental issues. Hence, Rwanda has one of the highest CE awareness rates among low-income countries (Špačková 2021). Consequently, the World Circular Economy Forum was hosted in Kigali, Rwanda, in 2022, and thus for the first time in a country of the Global South. Further, Rwanda has a National Circular Economy Action Plan and roadmap in place that is geared towards making Rwanda a carbon–neutral and climate resilient nation by the year 2050 (RoR 2022).

The objective of this research is to ascertain the level of awareness of Kigali's slum dwellers regarding CE-related concepts, to understand how they conceptualize CE, and to identify circular strategies already commonly practiced. The book chapter addresses the following research questions:

RQ1: What is the level of CE awareness and which perceptions prevail in Kigali's slums?

RQ2: What are the key CE-related practices in Kigali's slums?

This research provides a deeper understanding of the prevailing CE concepts for policy planning especially in awareness creation and community education for utilizing the CE potential in slums. Thus, the book chapter contributes to filling the research gap on socio-cultural aspects of CE in marginalized communities of the Global South.

2 Literature Review

While there is no standardized understanding of CE, most definitions share the emphasis on the upstream design of products and business models, a holistic and systematic approach, the prioritization of reuse and repair strategies over recycling, as well as designing out the idea of the end of product life (Dey and Iulo 2021; Goyal et al. 2018; *African Circular Economy Alliance*, n.d.; Ellen MacArthur Foundation, n.d.). The influential think tank Ellen MacArthur Foundation has distilled three main principles of circularity: eliminate waste and pollution, circulate products and materials at their highest value, and regenerate natural systems (Ellen MacArthur Foundation, n.d.). In terms of practical circular strategies, hierarchical frameworks of three to ten "Rs" are often cited, including the strategies recover, recycle, repurpose, remanufacture, refurbish, repair, reuse, reduce, rethink, and refuse (Potting et al. 2017). For the purpose of this study, we will focus on five R-strategies deemed relevant for informal settlements, namely recycle, reuse, refurbish, recover, and repair. The choice of the 5 Rs is justified by observational analysis on the common activities in the study areas. The interest in CE, which has lately grown in reaction to climate change and environmental harm, calls for a deeper analysis of how the idea and its process of transition are viewed in society (van Langen et al. 2021; Beehner 2023; Pál 2022).

Slums are "densely populated housing settlements" (D'Aquino et al. 2022, p. 1) associated with self-built housing, a lack of urban planning, poverty, vulnerability, and energy scarcity (D'Aquino et al. 2022; Ukpanyang et al. 2022). They are generally poorly integrated into the municipal infrastructure, and they grow quickly and without proper urban planning. Due to budget constraints, waste management is generally an underfunded area in many SSA countries (Allam and Jones 2018). Hence, solid waste management in slums poses a significant challenge and is associated with high costs, poor hygiene, lack of infrastructure, and environmental pollution (Allam and Jones 2018; D'Aquino et al. 2022; Ukpanyang et al. 2022,). It is estimated that less than 25% of the waste in informal settlements is collected on average, with incineration and landfilling prevailing (Ukpanyang et al. 2022).

Previous research on CE initiatives in slum areas has highlighted the importance of consistently involving the local communities (Allam and Jones 2018; Couder and Kibuthu 2020; D'Aquino et al. 2022). CE projects heavily rely on behavioural changes of the stakeholders, and in slum areas, peer-to-peer recommendations are critical for building trust in new products (Couder and Kibuthu 2020). CE projects in slum areas can only succeed if supported by the informal sector (Dey and Iulo 2021). Negative repercussions on informal workers who have hitherto been taking care of waste processing must be avoided to create a CE that is not only environmentally beneficial but also socially equitable (Ukpanyang et al. 2022; Gutberlet et al. 2017). Because the involvement of local inhabitants is important for CE viability, their perceptions and awareness of CE are key success factors.

Debrah et al. (2022) identify some of the obstacles to CE transition in SSA, such as little public knowledge on waste management initiatives as well as instructional programmes, a lack of political will, and insufficient financing and national waste management policies. Geisendorf and Pietrulla (2018) attribute the lack of openness to CE by various stakeholders to the CE discourse's inadequate incorporation of specific social and cultural aspects, such as housing, food systems, labour conditions and workers' rights, and international law. Similarly, Rotolo et al. (2022) found that social scepticism and lack of information about what CE embraces are still perceived among citizens in Argentina. Ogutu et al. (2023) state that businesses in Rwanda are less likely to adopt CE and other sustainability policies when there is insufficient government assistance, such as a lack of funding and capacity building to create new business models, an unclear regulatory environment, and insufficient stakeholder engagements.

Consumers play various roles within the CE, including purchasers, maintainers, repairers, sellers, sharers, collaborators, and waste discarders (Camacho-Otero et al. 2018). However, consumer awareness levels are still low. For example, a study by Sijtsema et al. (2020) discovered that Dutch consumers do not have a clear understanding of what CE means exactly. Yet, consumers have a much more active role in a CE than in a linear economy and are required to adapt many of their established habits and values (Sijtsema et al. 2020). In a study on CE consumer perceptions in the food domain in Portugal, Sousa et al. (2021) discovered significant misperceptions and uncertainties about the topic of food upcycling. Therefore, Sousa et al. (2021) conclude that awareness raising among consumers and more research about consumer perceptions are important steps for a successful shift towards CE. Thus, the current situation is characterized by a lack of consumer awareness and acceptance of the CE concept, which is particularly true in developing countries (Camacho-Otero et al. 2018; Kirchherr et al. 2017).

Awareness, perception, and practice are integral components influencing human behaviour and decision-making. At the forefront, awareness pertains to an individual's knowledge or consciousness of a particular subject or phenomenon (Klatzky 1984). It serves as the foundation for understanding, as individuals must be cognizant of a concept before forming perceptions or engaging in related practices. Perception, on the other hand, involves the interpretation and understanding of information, shaping an individual's attitudes and beliefs towards a specific matter (Chalmers 1996). As a cognitive filter, perception plays a vital role in mediating the transformation of awareness into action. Finally, the practice encompasses the actual behavioural implementation of knowledge and perceptions.

3 Methodology

3.1 Study Objective

Little is known about consumers' and practitioners' perceptions of CE in Rwanda in general, and even less about the perceptions and prevalence of CE in slum areas. This study aims to enhance our understanding of consumer awareness, attitudes, and perceptions in slum areas of Kigali City, Rwanda. The study focuses on both the general notion of the CE and specific CE practices and consumer perceptions towards CE among slum dwellers in Kigali. Through 15 key informant interviews and 12 focus group discussions on these topics, we explore how consumers perceive the CE at both abstract and concrete levels. The objective is to contribute to the emerging field of CE research by uncovering consumer perspectives and identifying potential entry points for interventions, research, and communication activities that can foster consumer interest and engagement.

3.2 Study Area

Administratively, Rwanda is divided into five provinces, namely; Eastern, Western, Northern, Southern, and Kigali (the Central Province), which is the capital city of Rwanda. The five provinces of Rwanda are composed of 30 districts. Each district is further divided into sectors, cells, and villages. Kigali City is comprised of three districts (Nyarugenge, Gasabo, and Kicukiro), as shown in Fig. 1. Within Kigali, Gasabo is the largest district by geographical area at 429.3 km^2, followed by Kicukiro with 166.7 km^2 and Nyarugenge with 134 km^2. Kigali lies at an altitude of 1,567 m and an area of 730 km^2, and it is located at a latitude of 10°58′S and a longitude of 30°07′E (Singirankabo and Iyamuremye 2022). Kigali receives an average of 950 mm of rainfall annually, with an annual average high and low temperature of 27 °C and 6 °C respectively (Bonfils 2012). Currently, the population of Kigali City is estimated to be 1.5 million people. As the commercial and administrative centre of Rwanda, Kigali City holds a crucial position in the country's economic development. The city generates over 50% of the national GDP, which highlights the significance of the city in the country's 11% GDP growth (CoK 2012). However, all three districts of Kigali have areas with slums.

This research was conducted in the urban slums of Kigali City. In Nyarugenge, data was collected from Gitega and Muhima sectors, from Umucyo, Sangwa, and Ituze villages. In Gasabo, data was collected in two sectors, namely Kimironko and Kinyinya in Nyagatovu, Giheka, and Muhororo villages. In Kicukiro, data was collected in Gatenga sector, mainly from Isangano and Amahoro villages. Unlike Kicukiro and Nyarugenge Districts, Gasabo district is characterized by a mixture of high mountains with an average altitude of 1,800 m mainly located in the rural zone, sloping basins and valleys. Kigali's city centre is situated in Nyarugenge district and

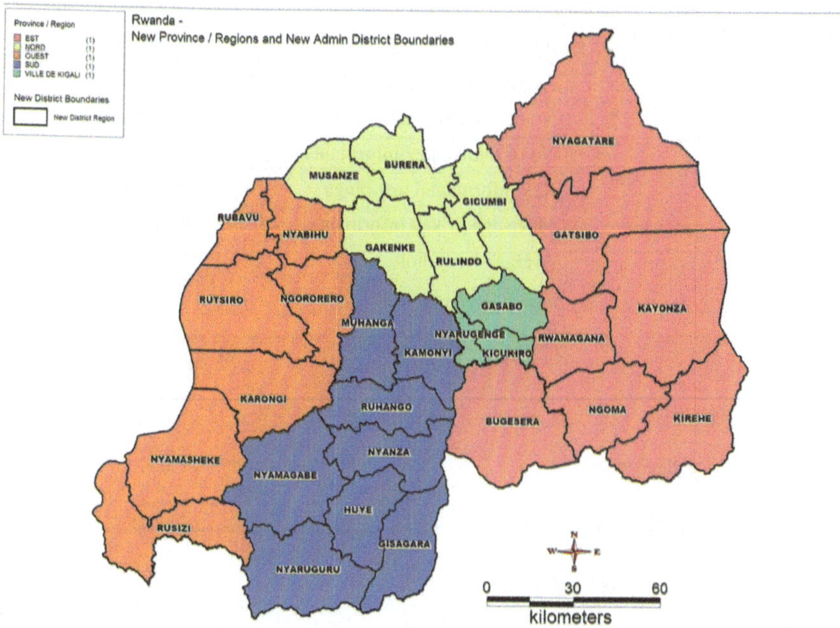

Fig. 1 Rwanda districts map (Reproduced from Government of Rwanda 2009)

it contains most of the city's businesses. Kicukiro is home to a thriving market and a number of Non-Government Organizations.

3.3 Research Design and Samples

This study adopted an exploratory research design. Two different data collection methods were triangulated, namely key informant interviews (KIIs) and focus group discussions (FGDs). The triangulation with two different data collection methods served to corroborate the findings and to increase their validity. Eight villages constituting some of the major slums of Kigali City were selected for this study and at least one cell was identified in each of the three districts. A total of five cells were included in this study (i.e., from Gasabo, Nyarugenge, and Kicukiro). A total of 122 respondents were engaged in this study. Fifteen community leaders were purposively selected as key informants (KIs), with five from each district. Key informants are the leaders of villages or cells and were selected purposively because they play an important role in coordinating their communities and are also knowledgeable in economic activities and practices. In addition, they were helpful in recruiting participants for the FGDs. Table 1 gives an overview of KI's biodata, and Fig. 2 indicates their level of education.

Table 1 Key informants' information

Information	Value
Average population of villages	811
No of male key informants	9
No of female key informants	6
Key informants' age range	19–59 years
Mean age of key informants	45 years
Average household size	6

Source Own research

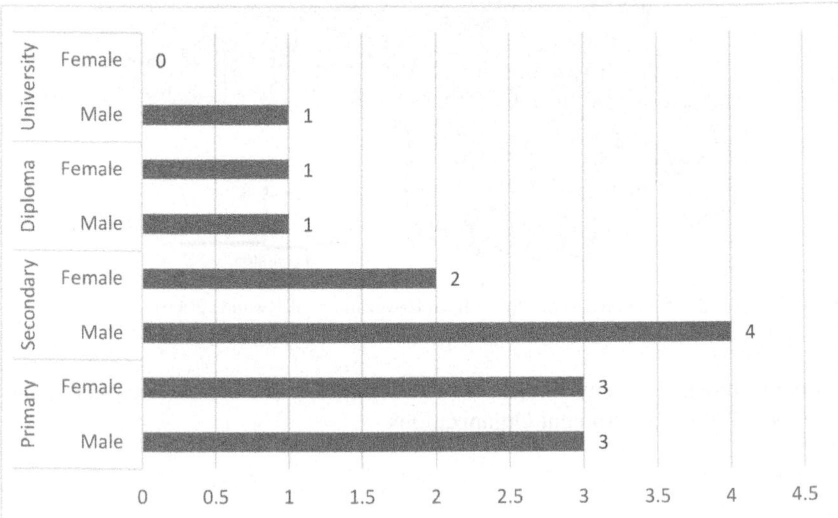

Fig. 2 Key informants' level of education. *Source* Own research

A total of 107 participants were interviewed through FGDs. The focus groups (FGs) were divided according to gender and marital status and four categories including men, women, and youth (boys and girls) were formed for the purpose of this study. One FGD per category was led in each of the three districts of Kigali City, leading to a total of 12 FGDs with seven to ten participants each. An overview of the conducted FGDs is provided in Table 2. Participants' codes can be read as such: GYF_P5 indicates the fifth young female participant from Gasabo. Respondents represented typical professions in their neighbourhoods, i.e., small business owners, tailors, cleaners, housemaids, motor taxi drivers, security guards, shoemakers and repairers, hairdressers, mobile money agents, farmers, housemaids, as well as sexual workers. FG participants were recruited through the village leaders, who also served as KIs.

Table 2 Overview of focus group discussions

FG code	District	Sector, cell, village	Gender	Age	No of participants
GYF	Gasabo	Kinyinya, Kagugu, Muhororo	Female	18–31 years	7
GYM	Gasabo	Kinyinya, Kagugu, Muhororo	Male	21–28 years	8
GOF	Gasabo	Kinyinya, Kagugu, Giheka	Female	30–44 years	9
GOM	Gasabo	Kimironko, Nyagatovu, Urugwiro	Male	38–58 years	7
KYF	Kicukiro	Gatenga, Gatenga, Amahoro	Female	22–34 years	8
KYM	Kicukiro	Gatenga, Gatenga, Amahoro	Male	19–38 years	10
KOF	Kicukiro	Gatenga, Gatenga, Isangano	Female	25–63 years	10
KOM	Kicukiro	Gatenga, Gatenga, Isangano	Male	25–73 years	7
NYF	Nyarugenge	Gitega, Kigarama, Umucyo	Female	20–33 years	10
NYM	Nyarugenge	Gitega, Kigarama, Umucyo	Male	18–37 years	10
NOF	Nyarugenge	Gitega, Kigarama, Umucyo	Female	27–69 years old	10
NOM	Nyarugenge	Gitega, Kigarama, Umucyo	Male	37–68 years old	9

Source Own research

3.4 Data Analysis Method

The questionnaire for KIIs included 5 and 3-point Likert scale rating questions. On awareness, the respondents were asked if they were aware, neutral, or not aware of key concepts related to CE. To capture the knowledge, attitudes, and behaviour towards

CE, the respondents were asked to indicate how strongly they agreed or disagreed on a 5-point Likert scale on a given statement. For data analysis, the Likert scale was coded from 1 to 0 respectively and the mean level of agreement was calculated. The data from the FGDs was transcribed and translated to English. After conducting the 12 FGDs, a certain level of theoretical saturation was reached; meaning that the later interviews yielded only limited new findings (Mayring 2007). The 12 FGDs from the three districts, separated by gender and age groups, allow for comparison between age, gender, and districts.

The data was interpreted using the qualitative content analysis (QCA) approach, a systematic, comprehensible, and replicable method to analyse qualitative data within its context (Mayring 2010; Schreier 2012). The data was coded using the web-based tool QCAMap developed by Mayring and Fenzl (2020). An inductive coding strategy was employed to uncover the relevant themes directly from the data, and analytic memos were written in parallel as suggested by Miles et al. (2020). Codes were subsequently grouped into overarching categories. QCA leads to a deep understanding of the data, but generalizability cannot be inferred from qualitative research (Mayring 2016).

4 Findings

4.1 CE Awareness and Perceptions

The FGDs revealed that the term CE is not very well known in Kigali's slum areas. When asked directly, participants of nine FGs largely agreed that they are not aware of the benefits of CE and lack incentives. Respondent NYF_P3 clearly expressed: "No idea on what circular economy means and how it can improve our welfare." Overall, participants from the four FGs in Nyarugenge expressed the least awareness of CE. Among the three study districts, slums in Nyarugenge are economically the weakest. There might be a relationship between CE awareness and socioeconomic situation.

The KIIs confirmed the pattern as demonstrated in Fig. 3. The KIs are strongly aware of general concepts connected to CE like economy, environment, as well as health and nutrition. These topics are of universal concern in slum areas. Respondents also demonstrated strong awareness regarding the five most prominent circular R-strategies (Recycle, Reuse, Refurbish, Recover, and Repair). Thirteen out of 15 KIs were aware of the five Rs, while only three had no opinion or no adequate knowledge of the same. However, awareness regarding more specific topics like circularity, resources, innovation, and behaviour was less developed. For example, 9 out of 15 respondents indicated awareness that resources (goods, energy, products) are a key component in the CE discourse while six had no clear awareness of the same.

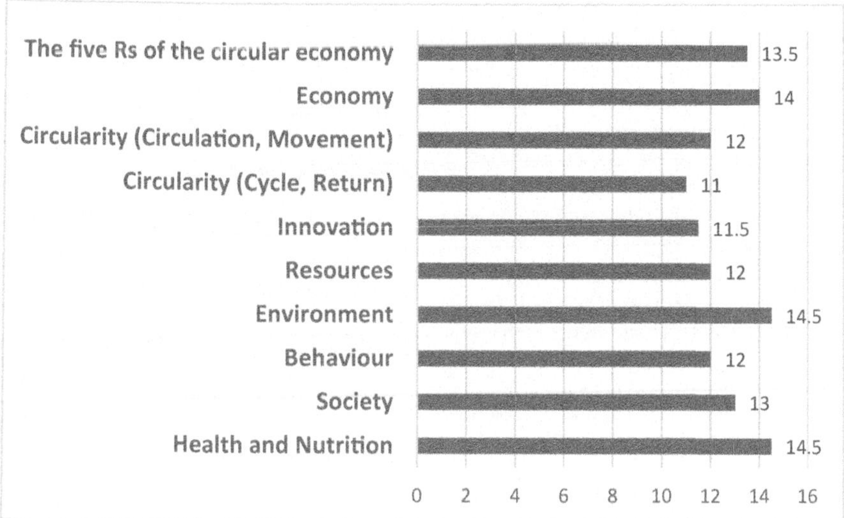

Fig. 3 Key informants' self-indicated level of awareness of CE-related concepts. *Source* Own research

When probed for circular practices with concrete examples, the FG participants largely displayed at least a minimum understanding of the concept, though not necessarily under the term CE. All FG participants could relate to the concept of circularity and name examples of circular practices from different aspects of everyday life. The strong reliance on the government as actor became apparent in a statement by NYF_P8: "*We think the government will find other ways to use the wastes in some useful ways.*" This indicates that she generally sees CE as somebody else's business.

Only two FG participants clearly articulated that they had a positive attitude towards CE. When referring to the village inhabitants as actors, the FG participants mostly highlighted shortcomings and a lack of incentives, but also pragmatic circular strategies like reusing and repurposing. In line with these findings, the KIIs revealed a pragmatic attitude towards CE, as illustrated in Fig. 4. The statements "It is better to buy used stuff than new ones" and "I would rather repair a broken item than buy a new one" received the highest approval rates. Statements that assign environmental responsibility to business owners also achieved fairly high approval rates. Two and seven KIs strongly agreed and agreed respectively that separating organic from inorganic garbage is a waste of time since at the final disposal site, the materials will be mixed again. However, three, one, and two were indifferent, disagreed, and strongly disagreed on the same. This indicates that the knowledge about and attitude towards waste separation is varied in the Kigali slum areas.

All FG participants from the three areas were in agreement that linear practices predominate. For example, GOF_P7 replied: "*In Giheka village, there are no significant businesses related to waste. I think we are still practicing linear economy.*" KYM_P6 related it to informality: "*Predominantly, there is linear economy and all*

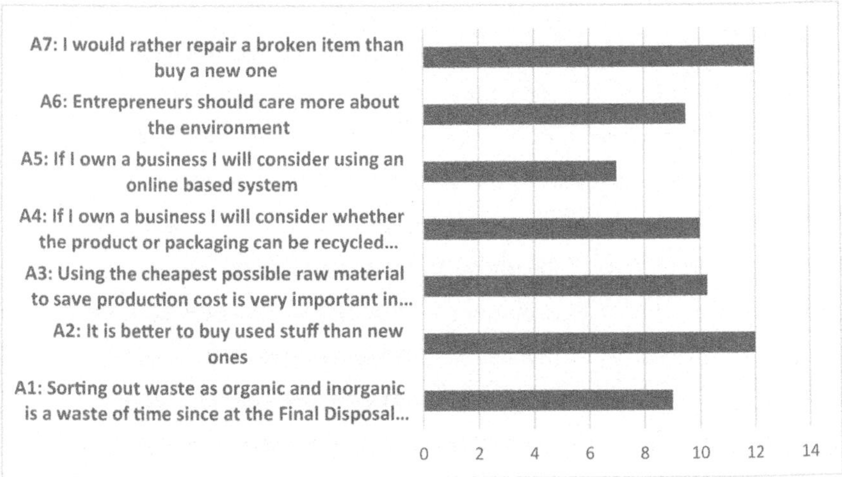

Fig. 4 KI's level of agreement to CE-attitude related statements. *Source* Own research

activities mentioned above are done or pursued informally and by default." NYF_P10 similarly highlighted the informality of the second-hand market: "*There are some people who buy old stuff and they find a small market for them, such as old shoes and clothes, but this is just done informally and on a very small scale.*" Notably, most of the FG participants emphasized the lack of circular *businesses*, which implies that their reluctance stems from the term *economy* in CE. The concepts of circularity and circular practices, however, are well-established in an informal way, though less related to businesses.

Business-oriented circular methods, such as taking part in second-hand markets, are seen as being mostly carried out by outsiders. In nine FGs, the villagers discussed that they are unaware of the purposes for which secondary materials are used, as these materials are purchased by outsiders. Therefore, many locals lack familiarity with the usage possibilities of secondary materials. For example, GYM_P4 responded: "*There are people who come in our village to buy used clothes, used plastic bottles, used jerricans and old metals but we don't know what they use them for.*" CE as something experienced in the professional realm was only mentioned by GOM_P3: "*The understanding of circular economy concept is informed by the day-to-day activities from our professions.*" Professions in this FG included construction workers, a businessman, a driver, a painter, and a cook.

The KIIs showed a slightly different picture. The majority of KIs agreed or strongly agreed that CE models support entrepreneurship. 11 and four respondents strongly agreed and agreed respectively that the adoption of circular business models is a key motivation for implementing CE. Twelve and three respondents strongly agreed and agreed respectively that CE promotes entrepreneurship. KIs, being village leaders with a comprehensive understanding of economic activities in their communities,

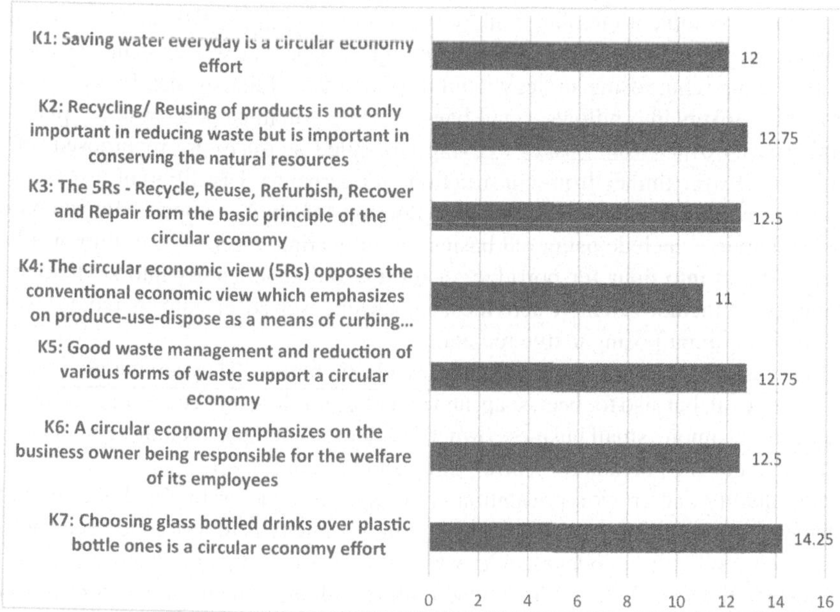

Fig. 5 KI's level of agreement to CE-knowledge related statements. *Source* Own research

are well-positioned to identify entrepreneurial opportunities. However, these opportunities were mentioned as motivations for embracing the CE, rather than being a present reality. This indicates that although slum residents perceive CE as a promising avenue for economic development, they have not yet fully capitalized on CE-related entrepreneurial prospects.

Recently, CE and circularity have developed into popular but weakly defined buzzwords. In order to ascertain the predominant understanding of CE in the informal settlements of Kigali, the KIIs asked respondents to indicate to which extent they agreed that certain practices are part of CE. The results are visualized in Fig. 5. Giving preference to glass bottled drinks, which are commonly refilled in Rwanda, achieved the highest level of acceptance as a circular practice. General waste avoidance, recycling and reusing materials for resource conservation, the 5R strategies as well as business owners' responsibility for employee welfare also achieved high agreement levels. This shows that CE is viewed as an overarching concept which encompasses economic, ecological, and social aspects of life.

4.2 *Circular Practices Common in Kigali's Slum Areas*

Seven major categories of circular strategies executed in the Kigali slum areas could be identified from the FGD and KII data, as summarized in Table 3. Repurposing is the most prominent circular strategy according to both the KIIs and the FGDs. Making envelopes for packaging from used office papers as well as cement sacs is the prevailing repurposing strategy in all three districts. Tailors generally repurpose garment leftovers into pillows, small bags, earrings, small carpets, or under plates. Similarly, leftovers from basket weaving (so-called agaseke) are repurposed into carpets. Leftover timber from construction and carpentry has different use cases: firewood, producing smaller furniture, or flooring for chicken coops. Minor repurposing activities include using old basins for cultivating vegetables, milling nearly expired bread into flour for porridge, or turning old shoe leather into keyholders. Repurposing includes creative activities, e.g. turning iron sheets into charcoal stoves or artistically repurposing textile remnants.

Reuse is a widespread strategy for empty plastic bottles. They are mainly reused for cooking oil, but also for beer, soap, and other liquids. Reusing boxes for packaging is customary among small business owners. Other reused items include cement sacs, household utensils, dreads for hairdressing, and broken floor tiles.

Soil quality and erosion prevention are of special concern in Kigali's hilly and densely populated landscape. The most common strategy is to turn food peels, non-consumed food, other biodegradable waste, and dust into organic manure to return nutrients to the soil. GYM_P4 mentioned filling soil into cement sacs to control soil erosion.

Participating in the market for secondary materials was discussed in 11 out of 12 FGDs, however portrayed as a strategy mainly employed by outsiders. The selling strategy is predominantly related to used metal, namely metal bars or iron metals, if described in more detail. Metal sales were mentioned in all FGs except GYF and NOF. Second-hand markets for clothes and shoes, bottles and jerricans, electronics as well as home essentials are of lesser importance. Astonishingly, the KIs did not mention the sale of secondary products and materials at all.

A further strategy is repairing broken items. It was mostly mentioned by male respondents and mainly related to shoes and clothes. A number of middle-aged, female respondents also mentioned the welding of saucepans. Producing animal feed from fruit and vegetable peels as well as non-consumed food leftovers was mentioned by participants of all four FGs in Gasabo. This strategy is applied in private households and by cooking chefs. It is only possible in Gasabo, since livestock keeping is banned within the city limits of Kigali, but Gasabo stretches into peri-urban, peasant areas where livestock keeping is widespread.

The KIIs also inquired about the respondents' level of agreement with statements about resource-saving behaviours according to the 5R strategies, which is visualized in Fig. 6. The results show that frugality and the appreciation of used products are firmly established in the study areas. Only renting things and carrying a refillable water bottle are rather unfamiliar.

Table 3 Prevalence of circular practices in the three slum areas according to FDG results

Circular practice	Mentioned by no of FGs	Mentioned by no of KIs
Repurposing	11	15
Paper into small envelopes	*9*	*15*
Garment and weaving leftovers for different uses	*4*	*1*
Carpentry waste for different uses	*3*	*1*
Basins/halved jerricans for cultivating vegetables	*1*	*6*
Old bread into flour porridge	*1*	*0*
Shoe leather to key holders	*1*	*0*
Reusing	11	3
Plastic bottles	*10*	2
Empty boxes for packaging	*4*	*0*
Jerricans	*3*	*0*
Cement sacs	2	*0*
Household utensils	*1*	*0*
Dreads for hair dressing	*1*	*0*
Broken floor tiles/bricks	*1*	*1*
Soil conservation measures	7	5
Organic manure	*7*	*5*
Cement sacs against soil erosion	*1*	*0*
Buying or selling used products	11	0
Metal	*10*	*0*
Clothes and shoes	*4*	*0*
Bottles and jerricans	*4*	*0*
Electronics	2	*0*
Home essentials	2	*0*
Repairing	5	2
Shoes	*4*	*1*
Clothes	2	*0*
Saucepan	*1*	*1*
Recycling	0	1
Metal	*0*	*1*
Turning food leftovers into animal feed	4	11

Source Own representation

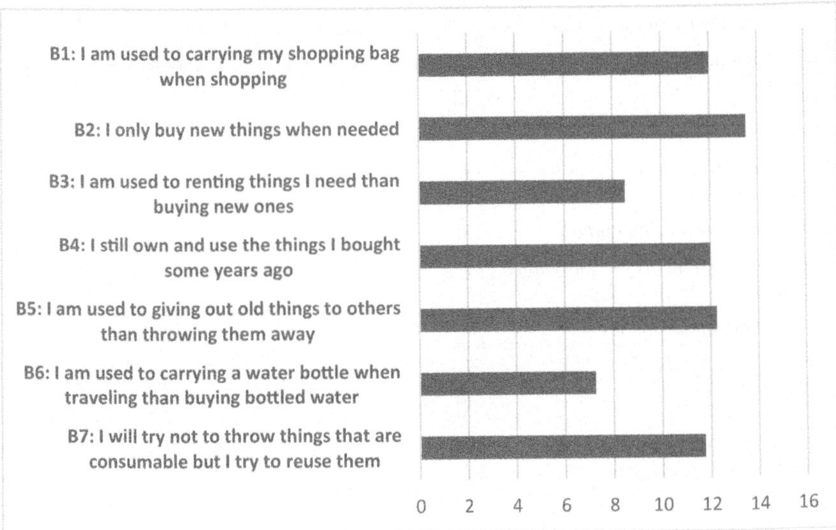

Fig. 6 KIs' level of agreement to CE-behaviour related statements. *Source* Own research

4.3 Perceived Benefits of CE

The FG participants largely agreed that financial benefits are the primary motive for participating in circular practices. Compared to the older FG participants, the younger generations are more likely to see business opportunities in circular activities. However, the theme of circular businesses does not come up in Nyarugenge, the area with the least CE awareness. In Gasabo and Kicukiro, circular activities are seen as opportunities for additional income and employment. FOM_P5 explicitly mentioned the positive social effects of creating employment opportunities: "*less robbery and prostitution.*" The perceived circular employment opportunities include jobs offered by waste management companies, whereby it must be mentioned that waste collection companies do not necessarily engage in circular R-strategies. Participants of the Nyarugenge FGs mostly placed emphasis on job opportunities rather than business opportunities, i.e., CE is less related to entrepreneurship in this district.

Interestingly, KII results place more emphasis on entrepreneurial opportunities than on other economic incentives for engaging in CE, as visualized in Fig. 7. All the KIs noted that the CE played a vital role in their villages through job creation and earning the participants some income. The statement "recycled products are cheaper" achieved the least level of agreement, which indicates that the recycling strategy is of minor importance to slum dwellers. Recycling encompasses the mechanical or chemical breakdown of materials into their smallest components to create like-new raw materials. However, this industrial process is expensive and beyond the control of consumers. Consequently, slum residents are frequently unaware of the proportion of recycled content in products, and cost savings from recycling are often unattainable.

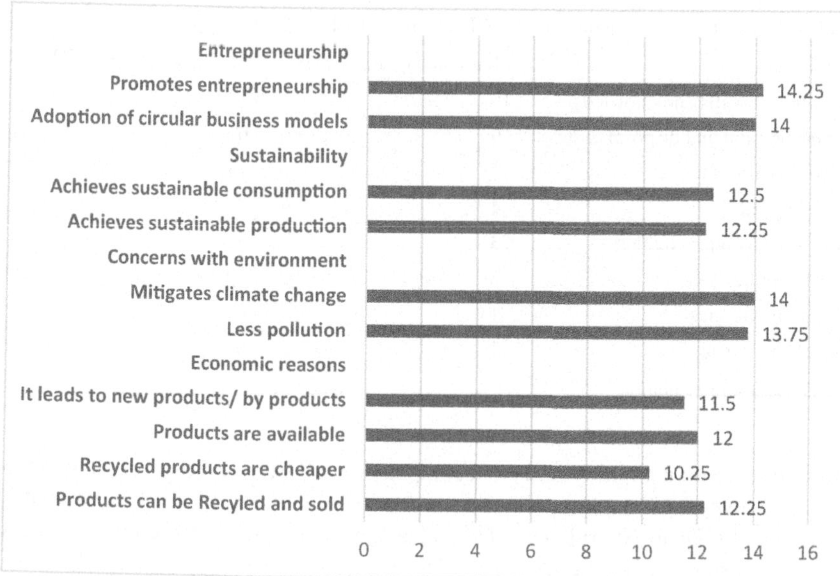

Fig. 7 Motivations for CE-related activities according to the 15 KIIs. *Source* Own research

Increased cleanliness of the surroundings is the second most important benefit of CE. In Rwanda, cleanliness is highly valued, and the government has strict regulations in this field. Participants of nine FGs cited hygienic concerns as a motivation. However, as highlighted by the KII results and illustrated by a quote from NYF_P5, the benefit of cleanliness is not only related to circular R-strategies but also to waste collection in general, which is a service the villagers pay for: "*We are only throwing away any unused materials to the garbage and even pay the companies to take the waste as it can bring flies and poor hygiene if they are kept near the homes.*" The KIIs revealed a similarly strong emphasis on environmental concerns and sustainability in general.

4.4 Challenges and Barriers in CE Implementation

FG participants in Nyarugenge discussed CE as the most problematic. The prevalence of key barriers to CE implementation is indicated in Table 4. The most significant challenge, discussed in ten out of 12 FGs, is insufficient skills and knowledge. For example, FG NYM_P1 mentioned: "*We don't have enough skills to sort waste and make it useful.*" Likewise, 13 out of 15 KIs highlighted insufficient waste sorting and valorization skills and knowledge as a key challenge for CE in their area.

On whether they had any waste materials in the surveyed villages, all the KIIs noted that they paid between RWF 2,000 and 3,500 for households and RWF 5,000

Table 4 Challenges in implementing the CE according to the FGD findings

Challenge	Mentioned by no of FGs	Mentioned by no of KIs
Insufficient skills and knowledge	10	13
Transportation and disposal costs	9	0
Waste separation	7	13
Poor hygiene	4	1
No clear secondary market	3	9
Volatile prices	3	10
Inappropriate sacs	1	0
Hazardous waste	1	0

Source Own representation

for businesses on a monthly basis to a company to collect their waste. High logistics and waste disposal costs were also discussed as a key challenge in nine FGDs, notably by all four FGDs in Nyarugenge. They clearly pose a barrier to trading secondary raw materials and taking them to specialized places for further processing. However, in this context, high transportation and disposal costs mostly referred to the fees charged by waste collection companies. KOF_P5 phrased it as follows: "*There is a high cost charged by waste companies who come to take wastes from households and homes.*" Hence, local reuse and repurposing strategies can present an opportunity to mitigate high waste collection costs. This, in turn, requires apt product design for durability and flexibility.

As indicated above, the issue of separating waste properly is strongly related to the skills gap. GYM_P1 put it as follows: "*We are not able to separate degradable and non-degradable wastes.*" KOF_P7 related the challenge to the lack of clear incentives: "*Sorting of wastes is a challenge in the village and there is no incentive for households to separate or sort waste.*" Pragmatic issues also hinder the correct separation of waste: the lack of appropriate sacs was mentioned by GYF_P2 and KOF_P10 highlighted the dangers of hazardous waste: "*Wastes from electronic materials have no use and they are problematic for us.*" In addition to the pragmatic issues, separating waste is not sufficiently incentivized. Even though Rwanda is in the process of implementing a policy of separating organic from non-organic waste, this is poorly implemented in the slum areas because the efforts for sorting garbage are not rewarded and the relevant skills are deficient.

Poor hygiene is related to the separating difficulties with insufficient sanitation being a general problem in poor neighbourhoods. Hygiene is most challenging in Nyarugenge due to poor access roads and a very dense population which makes it extremely difficult to manage solid waste, especially among crowded households. CE may have the potential to alleviate this situation, although this is not yet realized.

The practice of selling secondary materials and products presents market-related challenges. In the first place, it is difficult to find a market for secondary materials in the study areas. GYM_P8 explained: "*There is no exact place to sell our products from waste.*" It would be interesting to compare these results from Kigali with results from slum areas in other countries, since the highly regulated nature of commerce in Rwanda is unfavourable towards informal street merchants. Again, most used products are usually sold at low prices, since no competing use cases are available. KOM_P7 described the unstable nature of prices for secondary materials as follows: "*There is no standard price for products that are classified as waste. Only negotiated prices with informal buyers prevail.*" This highlights the vastly informal nature of the second-hand market, which was confirmed by FGs GYM and GOF as well as 10 KIs.

5 Discussion

While awareness of the term CE is low in the three slum areas of Kigali, the key principles of circularity (reusing products and materials, eliminating waste and pollution, regenerating natural resources) are part of their culture and habitual practices. However, CE is mainly perceived as a strategy handled by the government or by "outsiders," rather than embraced by the slum inhabitants as a means to improve their livelihoods. In this respect, the term circular *economy* can be deterring since *economy* connotes formality, power, and wealth. It might be more expedient to emphasize the concept of *circularity* and the concrete *R-strategies* that are already widely practiced, like reusing, repurposing, and repairing. The respondents in the study areas identified job creation and income generation as a significant benefit of the CE. This expectation aligns with previous research by Allam and Jones (2018), who highlighted the potential for creating 3,400 direct permanent jobs and generating US$2,380,000 annually through household waste recycling in Cotonou, Benin. Additionally, Korsunova et al. (2022) observed that economic necessity predominantly drove CE practices in informal sectors in Brazil, India, and Tanzania.

The KIIs and FGDs reveal that a number of circular activities exist among the slum dwellers. Circular practices are well-established in different areas of professional and private life in the slum areas. They are predominantly located in the informal as well as non-monetized economy, whereas more formalized circular business models play a minor role. Widespread practices are: In the surveyed slums, common widespread practices of the circular economy include paper recycling into small envelopes and using food waste as animal feed. Other activities involve repurposing plastic jerricans for vegetable cultivation and converting food peels into organic manure. While recycling and food-related circular practices show successful adoption, other activities like reusing beverage bottles, carpentry waste, and garment remnants remain minimal. These findings are in line with the potential for circular economy practices in recycling and the food sector, as supported by Schröder et al. (2019).

However, there is still potential to scale circularity. The high costs for waste disposal are a burden for the inhabitants of slums, and unsustainable waste management adversely affects health and environment. Hence, scaling circular practices and establishing more formal circular businesses and second-hand markets are promising strategies. Yet, a lack of skills and knowledge has been identified as the key barrier to implementing CE. This can be addressed with targeted trainings and awareness-raising campaigns. Measures should instil hands-on skills. The catalogue of easily replicable do-it-yourself instructions by the startup Hack Your Shack (2022) can be an inspiration. Village leaders served as KIs for this research and showed more awareness of CE than the general population. Thus, they can become key persons in multiplying CE-related skills.

This research also demonstrates that selling secondary materials is problematic in Kigali's slums. Prices are low and unstable, and clear sales points are unavailable. Establishing regulated second-hand markets can help to promote reusing and repurposing items. The results show that renting items as well as internet-based business models are not yet widespread. The Rwandan startup Wastezon (2022), however, demonstrates that digital platforms can be used to leverage refurbishing and C2C sales of used electronics in Rwanda. Similar models might be extended to Kigali's slum areas to serve a wider population. However, cycling of secondary materials should be restricted locally. Fevrier (2022) cautions that importing used products from the Global North for processing in informal settlements of the Global South can be a form of racialized capitalism.

In order to harness the sustainability potential of the CE, it must be ensured that products disseminated in slum areas adhere to circular design principles to guarantee durability, safety, and flexible usability. For this to be successful, a comprehensive nationwide circular design policy must be implemented, encompassing measures beyond the prohibition of single-use plastic. Such policy measures should include provisions for the right to repairability, extended producer responsibility, the polluter pays principle, and the introduction of a deposit scheme for packaging materials. By fostering innovation, creating product standards, and delivering education and awareness programmes, the Rwandan government and local stakeholders can collaborate to provide leadership, funding, and coordination while assisting the transition of the slum areas to a more *circular* economy.

6 Conclusion

This chapter has provided an overview of CE awareness, perceptions, and practices in Kigali's slum areas based on descriptive qualitative research. The culture of circularity is already firmly established in the informal sector and non-monetized economy of Kigali's slums. This culture should be embraced and fostered. While the term circular *economy* is regarded somewhat reservedly, pragmatic circularity strategies like reusing and repurposing items are widely practiced. Targeted measures can contribute to the scaling of circularity in slums for the benefits of the inhabitants. This includes upskilling inhabitants, establishing second-hand markets, and regulating product design.

This research provides interesting insights for policymakers and the academic community. It contributes to the literature on CE awareness and perceptions as well as the literature on sustainable development in informal settlements. However, the validity of the findings has certain limits. Data was collected using KIIs and FGDs rather than through direct observation. Hence, the bias of social acceptability cannot be excluded and the actual prevalence of circular practices cannot be quantified. As the research design is descriptive and qualitative with a small sample size of 15 KIs and 12 FGDs with a total of 107 participants, the findings cannot be considered generalizable.

This study can serve as a starting point for further research. Observational studies and longitudinal studies of how prevalent circular practices are would make the evidence basis more profound. A further promising research avenue is action research projects which implement regulated second-hand markets or circular design policies for products, and document their impact on the overall economic structure of the informal settlements as well as potential environmental benefits.

Acknowledgements This research was funded by the DAAD with the financial means of the German Federal Ministry for Economic Cooperation and Development (BMZ). However, the opinions expressed here are those of the authors and do not reflect the position of the funding agency. The authors do acknowledge the guidance from Prof. Dr. Thomas Bayer, the Lead Principal Investigator (PI) for Advancing Sustainable Economic Development in Rwanda through Circular Agro-Economy (Rwanda-CAE) Project through which the research funding was secured. Finally, the authors recognize the collaboration with the community leaders and individuals in the selected urban slums where data was collected.

References

African Circular Economy Alliance (n.d.). Retrieved 27.06.2022 from www.aceaafrica.org

Allam Z, Jones DS (2018) Towards a circular economy: a case study of waste conversion into housing units in Cotonou, Benin. Urban Sci 2(4):118. https://www.mdpi.com/2413-8851/2/4/118

Beehner CG (2023) System leadership for overcoming the impossibilities of a circular economy. In: Lehmann H, Hinske C, de Margerie V, Slaveikova Nikolova A (eds) The impossibilities of the circular economy. Routledge, pp 156–166. https://doi.org/10.4324/9781003244196-18

Bonfils S (2012) Trend analysis of the mean annual temperature in Rwanda during the last fifty two years. J Environ Prot 3(6):538–551. https://doi.org/10.4236/jep.2012.36065

Camacho-Otero J, Boks C, Pettersen IN (2018) Consumption in the circular economy: a literature review. Sustainability 10(8):2758. https://doi.org/10.3390/su10082758

Chalmers DJ (1996) The conscious mind. Oxford University Press, In Search of a Fundamental Theory

CoK (2012) The city of Kigali development plan (CKDP) 2012/13–2017/18

Corvellec H, Stowell AF, Johansson N (2022) Critiques of the circular economy. J Ind Ecol 26(2):421–432. https://doi.org/10.1111/jiec.13187

Couder L, Kibuthu S (2020) Unlocking the circular economy potential to tackle the sanitation challenge. Field Act Sci Rep 22:72–77

D'Aquino CA, Pereira BA, Sawatani TF, de Moura SC, Tagima A, Ferrarese JCBB, Santos SC, Sauer IL (2022) Biogas potential from slums as a sustainable and resilient route for renewable energy diffusion in urban areas and organic waste management in vulnerable communities in São Paulo. Sustainability 14(12):7016. https://www.mdpi.com/2071-1050/14/12/7016

Debrah JK, Teye GK, Dinis MAP (2022) Barriers and challenges to waste management hindering the circular economy in Sub-Saharan Africa. Urban Sci 6(3):57. https://doi.org/10.3390/urbansci6030057

Dey S, Iulo LD (2021) The circular economy of Dharavi. ARCC J Archit Res 18(2)

Ellen MacArthur Foundation (n.d.) What is a circular economy? Ellen MacArthur Foundation. https://ellenmacarthurfoundation.org/topics/circular-economy-introduction/overview. Accessed 24 October 2022

Esken B, Franco-García M-L, Fisscher OAM (2018) CSR perception as a signpost for circular economy. Manag Res Rev 41(5):586–604. https://doi.org/10.1108/MRR-02-2018-0054

Fevrier K (2022) Informal waste recycling economies in the global south and the chimera of green capitalism. Antipode 54(5). https://doi.org/10.1111/anti.12841

Geisendorf S, Pietrulla F (2018) The circular economy and circular economic concepts—a literature analysis and redefinition. Thunderbird Int Bus Rev 60(5):771–782. https://doi.org/10.1002/tie.21924

Goyal S, Esposito M, Kapoor A (2018) Circular economy business models in developing economies: lessons from India on reduce, recycle, and reuse paradigms. Thunderbird Int Bus Rev 60(5):729–740

Gutberlet J, Carenzo S, Kain J-H, Martiniano de Azevedo AM (2017) Waste picker organizations and their contribution to the circular economy: two case studies from a global south perspective. Resources 6(52):1–12. https://doi.org/10.2290/resources6040052

Gregson N, Crang M, Fuller S, Holmes H (2015) Interrogating the circular economy: the moral economy of resource recovery in the EU. Econ Soc 44(2):218–243. https://doi.org/10.1080/03085147.2015.1013353

Hack Your Shack (2022) Hack Your Shack: Sharing inovation - Improving lives. https://hackyourshack.org/. Accessed 27 March 2023

Klatzky RL (1984) Memory and awareness: an information-processing perspective. W. H. Freeman and Company

Kirchherr J, Reike D, Hekkert M (2017) Conceptualizing the circular economy: an analysis of 114 definitions. Resour Conserv Recycl 127:221–232. https://doi.org/10.1016/j.resconrec.2017.09.005

Korhonen J, Nuur C, Feldmann A, Birkie SE (2018) Circular economy as an essentially contested concept. J Clean Prod 175:544–552. https://doi.org/10.1016/j.jclepro.2017.12.111

Korsunova A, Halme M, Kourula A, Levänen J, Lima-Toivanen M (2022) Necessity-driven circular economy in low-income contexts: how informal sector practices retain value for circularity. Glob Environ Change 76(102573). https://doi.org/10.1016/j.gloenvcha.2022.102573

Lehmann H, Hinske C, de Margerie V, Slaveikova Nikolova A (eds) (2023) The impossibilities of the circular economy: separating aspirations from reality. Routledge

Mayring P (2007) On generalization in qualitatively oriented research. FQS Forum: Qual Soc Res 8(3)

Mayring P (2010) Qualitative Inhaltsanalyse, 11th rev ed. Beltz

Mayring P (2016) Einführung in die qualitative Sozialforschung, 6th ed. Beltz

Mayring P, Fenzl T (2020) QCAmap: a software for qualitative content analysis. https://www.qcamap.org/ui/home. Accessed 24 March 2023

Miles MB, Huberman AM, Saldana J (2020) Qualitative data analysis, 4th ed. SAGE

Ogutu MO, Akor J, Mulindwa MS, Heshima O, Nsengimana C (2023) Implementing circular economy and sustainability policies in Rwanda: experiences of Rwandan manufacturers with the plastic ban policy. Front Sustain 4:1092107. https://www.frontiersin.org/articles/10.3389/frsus.2023.1092107/full

Pál V (2022) Social and cultural aspects of circular economy: towards solidarity and inclusivity. In: Pál V (ed) Social and cultural aspects of the circular economy: toward solidarity and inclusivity, Ebook ed. Routledge. https://doi.org/10.4324/9781003255246-1

Potting J, Hekkert MP, Worrell E, Hanemaaijer A (2017) Circular economy: measuring innovation in the product chain. Planbureau voor de Leefomgeving (2544)

RoR, Ministry of Environment (2022) Rwanda national circular economy action plan and roadmap. RoR, Ministry of Environment, UNDP. https://www.environment.gov.rw/index.php?eID=dumpFile&t=f&f=58556&token=1efafef04395aa568ceac5346426c5d29864bced. Accessed 03 April 2023

Rotolo GC, Vassillo C, Rodriguez AA, Magnano L, Vaccaro MM, Civit B, Covacevich MS, Arena A, Ulgiati S (2022) Perception and awareness of circular economy options within sectors related to agriculture in Argentina. J Clean Product 373(133805). https://doi.org/10.1016/j.jclepro.2022.133805

Rwanda GO (2009) Rwanda Districts Map. In: Rwanda Districts Map.jpg (ed) Wikimedia Commons

Schreier M (2012) Qualitative content analysis in practice. SAGE.

Schröder P, Anantharaman M, Anggraeni K (2019) The circular economy and the global south: sustainable lifestyles and green industrial development. Routledge

Sijtsema SJ, Snoek HM, van Haaster-de Winter MA, Dagevos H (2020) Let's talk about circular economy: a qualitative exploration of consumer perceptions. Sustainability 12(286). https://doi.org/10.3390/su12010286

Singirankabo E, Iyamuremye E (2022) Modelling extreme rainfall events in Kigali city using generalized Pareto distribution. Meteorol Appl 29(4):e2076. https://doi.org/10.1002/met.2076

Sousa PM, Moreira MJ, de Moura AP, Lima RC, Cunha LM (2021) Consumer perception of the circular economy concept applied to the food domain: an exploratory approach. Sustainability 13(11340). https://doi.org/10.3390/su132011340

Špačková I (2021) The diffusion of the circular economy awareness across the world. Diploma's thesis (Mgr), Charles University. Prague

Ukpanyang D, Terrados-Cepeda J, Hermoso-Orzaez MJ (2022) Multi-criteria selection of waste-to-energy technologies for slum/informal settlements using the PROMETHEE technique: a case study of the greater Karu Urban area in Nigeria. Energies 15(10):3481

UN-Habitat (2020) World cities report 2020. https://unhabitat.org/sites/default/files/2020/10/wcr_2020_report.pdf. Accessed 13 July 2023

van Langen SK, Vassillo C, Ghisellini P, Restaino D, Passaro R, Ulgiati S (2021) Promoting circular economy transition: a study about perceptions and awareness by different stakeholders groups. J Clean Product 316(128166). https://doi.org/10.1016/j.jclepro.2021.128166

Van Weelden E, Mugge R, Bakker C (2016) Paving the way towards circular consumption: exploring consumer acceptance of refurbished mobile phones in the Dutch market. J Clean Prod 113:743–754. https://doi.org/10.1016/j.jclepro.2015.11.065

Vehmas K, Raudaskoski A, Heikkilä P, Harlin A, Mensonen A (2018) Consumer attitudes and communication in circular fashion. J Fash Market Manag Int J 22(3):286–300. https://doi.org/10.1108/JFMM-08-2017-0079

Wastezon (2022) Wastezon Ltd. Resell. Reuse. Remanufacture. Redefine the future of urban mining. https://wastezon.com/. Accessed 30 March 2023

Recycling from Construction and Demolition (C&D) Waste: Exploring the Scope of Circular Management in Constructing Slum Dwellings of Dhaka

Md. Nawrose Fatemi and Tahmina Rahman

Abstract The Construction Sector in Bangladesh has accelerated over the last few decades, particularly since Dhaka, the nation's capital, has been the focal point for major physical development-related activities such as the construction of new satellite towns, housing, apartments, and infrastructures. Dhaka's population is growing rapidly as a consequence of development initiatives, which led to a dramatic rise in the city's urbanization and a corresponding rise in the demand for housing facilities. Currently, the trend toward constructing new high-rise structures is continuously emerging, while existing low-rise structures are being demolished in and around Dhaka. Construction and Demolition (C&D) waste, in huge quantities, was, therefore, produced as a byproduct, then added to the waste stream, and ultimately headed for the landfills. Through content analysis of regulatory documents, conducting a questionnaire survey with slum dwellers, and direct observations, this research aims to explore the scopes and preferences of slum dwellers to use recycled C&D waste as construction products for slum houses in Dhaka. This study provides a substantial contribution to the concept of C&D waste reduction by proposing an alternate application for the construction or renovation of slum dwellings, as well as introducing novel approaches to circular waste management and improving the effectiveness of existing resource management practices.

1 Introduction

Urbanization in Dhaka, the capital of Bangladesh, has increased dramatically in recent decades (Kulsum and Moniruzzaman 2022). With a current population of 23.2 million, the City ranks as the seventh most populous in the world, up from a mere

Md. N. Fatemi (✉) · T. Rahman
Department of Architecture, University of Asia Pacific (UAP), Dhaka, Bangladesh
e-mail: nawrose@uap-bd.edu

T. Rahman
e-mail: tahminarahman@uap-bd.edu

S. A. Okyere et al. (eds.), *Urban Slums and Circular Economy Synergies in the Global South*, Advances in 21st Century Human Settlements,
https://doi.org/10.1007/978-981-99-9025-2_5

1.5 million in 1971, during the independence of Bangladesh (WPR 2023). Since the city's population has grown exponentially, the construction of new buildings along with the upgrading of the city's infrastructure has skyrocketed in recent years (Fatemi 2012). It is worth mentioning that according to the World Bank, around 10.3% of the country's GDP comes from the construction and infrastructure sectors, which play an important part in the economic development and growth of Bangladesh (World Bank 2023a). Currently, Dhaka has become one of the world's fastest-growing megacities, with a yearly population growth of roughly 4.2% and a population density of 23,234 per square kilometer within its entire territory of 300 square kilometers (WPR 2023). This population growth has put enormous pressure on the city's infrastructure, resulting in a need for new buildings and facilities to keep up with the expanding population. Moreover, Bangladesh's economy has been booming during the last decade, expanding by 6.9% annually on average (WB 2022). Dhaka, as the country's administrative and economic hub, has been observing an influx of investment as a result of this recent economic growth, with both local and foreign investors looking to capitalize on Bangladesh's expanding economy. Hence, this economic expansion has attracted more people to the city in pursuit of employment and better opportunities, increasing the need for housing and infrastructure. This has directed to a rise in the number of new high-rise buildings being constructed while existing low-rise buildings are being demolished in and around Dhaka (Fatemi 2012; Nipun et al. 2016; Zaman and Laing 2013). Since Construction and Demolition (C&D) waste was produced in such vast amounts as a byproduct, it was added to the waste stream and eventually destined for landfills.

An estimated 85% to 90% of all generated C&D waste ends up in landfills in Bangladesh (Hasan et al. 2022). Dhaka, like many other cities, struggles with excessive amounts of C&D waste (Datta et al. 2022). In Dhaka, a considerable amount of total municipal solid waste is comprised of materials from Construction and Demolition sites, with estimates ranging from 15 to 30% (Fatemi 2012). The City produced over 1.28 million tons of C&D waste in FY 2016–2017 (1.139 tons of demolition waste and 0.149 tons of construction waste) (Islam et al. 2019). In addition to being a severe environmental concern, the disposal of such waste in landfills depletes natural resources (Jamil and Jamal 2018). More than half of all C&D waste generated globally has been estimated to end up in landfills, creating a serious environmental problem that every country must tackle (Nagapan et al. 2012). As most of the landfills in developing countries are unsanitary and lack leachate and gas collecting systems, soil and water contamination, waste sludge, odors, and GHG emissions are common around those landfills (Ouda et al. 2018). Therefore, adopting the circular economy concept is a viable method to reduce C&D waste's negative environmental effects (Purchase et al. 2022). Research on the recycling and reuse of C&D waste adopting the circular economy concept in Bangladesh is still underdeveloped. Even, limited attention has been paid to the potential alternative uses of C&D waste, particularly, to facilitate the houses for urban slum dwellers in Dhaka via circular management of C&D waste, hence reducing the total cost of such housing. Therefore, this study investigates the scopes and preferences of slum dwellers to use recycled C&D waste as construction products for slum houses in Dhaka through content analysis of regulatory documents, conducting a questionnaire survey with slum dwellers, and direct observations. In addition to introducing novel approaches to circular waste management and enhancing the efficacy of existing construction management practices, this study makes a significant contribution to the concept of C&D waste reduction by recommending an alternative sustainable approach for the construction or renovation of slum dwellings.

2 Literature Review

2.1 *Circular Management of C&D Waste*

Waste from the construction and demolition sector accounts for about half of the total waste produced in several countries (Kibert 2016) such as New Zealand (Inglis 2007), Brazil (Favaretto et al. 2017), and Korea (Yang et al. 2015). However, C&D waste recycling has gained popularity as a way to accomplish the goals to promote resource efficiency and advance the cause of a circular economy (He and Yuan 2020; Shooshtarian et al. 2022). Mulder et al. (2007) suggested that C&D waste recycling might be possible if the construction industry adopted the circular management concept. Several components of C&D wastes are recyclable, including concrete, brick, wood, glass, metals, plastics, gypsum, excavated soils, asbestos, solvents, and so on (Dyer 2012; EPA 2021). Several countries in Asia initiated the adoption of different regulatory policies to minimize waste generation, recycling, and material recovery. Malaysia established several laws about how to deal with waste, including the Standard Specifications for Building Works (SBW) and the Solid Waste & Public Cleansing Management Act of 2007 (Nagapan et al. 2012). Singapore implemented the ISO 14000 Environmental Management System (EMS) for the construction industry to lessen the negative impact on the environment due to building construction and demolition (Ofori et al. 2000). In 2000, Japan passed the Construction Material Recycling Law, which is credited with helping to boost Japan's C&D waste recycling rate by over 33% between 1996 and 2008 (MoEJ 2019). Mandalay, Myanmar's former capital, has a plan to increase its recycling rate, especially for industrial and special wastes (which includes C&D waste) from its current rate of 35% to 80% between 2017 and 2030 (MCDC & ECD 2017). In Cambodia, Waste Management Strategy and Action Plan 2018–2035 are aimed to increase Phnom Penh's C&D waste collection to over 90% and recycling to over 85% by 2023 (PPCA 2018). In Vietnam, an initiative to increase the C&D waste recycling rate in Hanoi via the use of mobile crushers was recently advanced by the Hanoi Department of Construction (Qinghai 2017).

In South Asia countries, several initiatives have been taken to adopt 3R (reduce, reuse, and recycle) and SCP (Sustainable Consumption and Production) waste management policies, laws, and regulations both at national and local levels. In

India, for example, the Delhi government has mandated that all government building projects must employ secondary raw materials for at least 5% of the total materials used (Chandran 2016). However, waste collection, recycling, and material recovery are still widely practiced throughout these South Asian countries, mostly on an informal basis. The Solid Waste Management Rules 2021 in Bangladesh have just been released, marking the beginning of the country's 3R (Reduce, Reuse, Recycle) Strategies for solid waste management (UN 2021). The potential benefits of recycling the C&D waste in Dhaka extend well beyond the economic and environmental advantages, and might even give social benefits like more affordable housing options. Around 52% of Bangladesh's urban population as estimated by the World Bank, dwells in slums and other forms of informal settlement (WB 2023b). Due to a shortage of low-cost housing options, slums have expanded, leaving their residents in deplorable circumstances. The utilization of recycled C&D waste in the construction of low-cost houses might provide a sustainable solution to Dhaka's housing and construction waste-related challenges.

2.2 *Scope of Recycled C&D Waste in Constructing Slum Dwellings*

Current construction industry standards, particularly in developing countries like Bangladesh, do not allow for easy deconstruction or disassembly (Whittaker et al. 2021). After the 'soft stripping' of a building which entails the removal of all fixtures and fittings, the structure is demolished and later, the resulting C&D waste is destined for landfills (Sassi 2019). Moreover, as cited by Whittaker et al. (2021), the majority of the recycled C&D waste is used for low-value purposes, including pipe bedding or in pavement construction (e.g., road sub-base and base course). Several research works have examined the potential of reusing C&D waste as construction products: from recycled aggregates (Kore and Vyas 2016; Shahidan et al. 2017) to alternative innovative materials (e.g., plastic bottles as brick replacement) (Paihte et al. 2019) for constructing new buildings. Roychowdhury et al. (2020) corresponded that recycled and repurposed C&D waste may be employed for constructing new buildings once again, offsetting the need for certain natural building materials. According to Islam et al. (2019), Bangladesh earns $44.9 million annually from the recycling of brick (0.269 Mt) and concrete (0.576 Mt) waste. In 2018, over 6,500 tons of waste were produced per day in Dhaka, and that number is projected to rise to almost 8,500 tons by 2032 (Islam 2021). Dhaka North City Corporation recycled just 10% of these wastes in 2018 while aiming for 44% by 2025; at which point, the C&D waste collection rate is expected to rise to 695 tons per day and the recycling rate will improve to 336 tons per day (DNCC 2019).

In addition, the potential for utilizing numerous building components recovered from C&D waste to build urban slum houses is substantial, especially considering the vast quantity of waste generated through construction and demolition activities in a metropolis like Dhaka. The United Nations estimates that 227 million people (22% of the global total) in Central and South Asia require affordable and decent quality housing (UNSD 2019). Sanchez Rodriguez et al. (2018) also highlight the emphasis on adequate and affordable housing by several UN Sustainable Development Goals (SDGs), most notably Goal 11 (Sustainable Cities and Communities), which also contributes directly or indirectly to Goal 3, 5, 6, 10, 12, 15, and 16. Therefore, using recycled C&D waste in new and retrofit construction to make housing more affordable for slum dwellers can be one way to put the SDGs into practice. Nevertheless, due to inconsistencies in quality control, recycled C&D wastes may cause problems with the long-term performance of these dwellings in terms of material strength, durability, and safety (Liu et al. 2021). Even with these problems, the move toward circular management of C&D waste in redeveloping affordable houses in slums is likely to keep growing in the future years, as a growing number of individuals learn to value sustainable development and resource efficiency in the construction sectors.

3 Methodology

3.1 *Study Design*

Content analysis of six policy documents on waste management, conducting a questionnaire survey with slum dwellers in slum areas of different sizes in Dhaka, and direct observations were used to conduct this research. The content analysis was conducted to better comprehend the current state of C&D waste management and the potential for implementing circular C&D waste management. Regarding waste management, the Government of Bangladesh (GoB) has developed numerous policies, acts, regulations, and plans, as shown in Table 1.

For Bangladesh, to the best of the authors' knowledge, only these regulatory documents provide a comprehensive structure for waste management, with a focus on solid waste management policy at both the national and municipal levels. Solid Waste

Table 1 Regulatory Documents for Solid Waste Management in Bangladesh

Policy instruments	Year	Organization
Clean Dhaka Master Plan 2005–2015	2005	Dhaka City Corporation
National 3R Strategy for Waste Management 2010	2010	Department of Environment
National Environmental Policy 2013	2013	Department of Environment
National Environmental Policy 2018	2018	Department of Environment
New Clean Dhaka Master Plan 2018–2032	2019	Dhaka North City Corporation
Solid Waste Management Regulations 2021	2021	Ministry of Environment, Forest and Climate Change

Management Rules 2021 recently announced the nationwide 3R (Reduce, Reuse, Recycle) Strategies for solid waste management. The policy goals, scopes, gaps, and strategies regarding the reuse of C&D waste as construction products in building slum houses in Dhaka were the main findings from these regulatory documents. The focus in analyzing these policy instruments was on the several mentions of recycled C&D waste as construction products. A thorough review in these regulatory documents on all the steps for circular management of C&D waste (i.e., C&D waste material collection for reuse, construction with recycled C&D waste, use of infrastructure built with recycled C&D waste, End of Use and scope of further alternative Use) (Gálvez-Martos et al. 2018) was conducted. Consequently, these content analyses of the regulatory documents were categorized into four groups after, that is, those which mentioned "non-renewable construction products' use", "C&D waste collection", "C&D waste recycling", and "alternatives of C&D waste".

A questionnaire survey of 300 slum dwellers in 5 slum areas of Dhaka (Fig. 1) along with direct observations was also conducted for this research to understand their knowledge and preferences regarding the recycled C&D waste in building or retrofitting their houses. The questionnaire was designed using a Likert Scale with five points of agreement, likelihood, quality, value, and strength as the measuring criteria. Discussion with respondents who had experience using recycled C&D waste yielded the variables of the age of the house, the construction material collection procedure, and the familiarity with recycled C&D waste used in the analysis. The other three variables, i.e., the quality of recycled C&D waste, the cost of recycled C&D trash, and the durability of recycled C&D stated in the analysis, were tallied based on the respondents' preferences to utilize the recycled C&D waste in building or retrofitting their houses.

3.2 Study Setting and Data Collection

This research was conducted in five different slums in Dhaka, Bangladesh (Fig. 1). Three considerations guided this selection of study areas. First, about four million people live in Dhaka City's roughly 5,000 slums (CUS 2005; Rahman et al. 2014). Dhaka sees an annual influx of 300,000 to 400,000 people migrating from different parts of Bangladesh seeking better opportunities (Amjad 2021). These slum dwellers live in abject poverty, deplorable living conditions, and extremely overcrowded houses. Secondly, the Constitution of Bangladesh safeguards all of its citizens' right to adequate shelter. According to Article 15(a), access to "…the necessities of life, including food, clothing, shelter, education, and medical care" is a fundamental human right, and the government makes it a high priority to protect this right for all citizens (Ali et al. 2020). Correspondingly, The National Housing Policies of 1993, 2008, and 2017 all make similar commitments to provide all citizens with housing or accommodation of minimum quality to live (Ahmed 2019). More importantly, to the best of the authors' knowledge, no other research in Bangladesh has attempted to address both the issues, the recycling and reusing of C&D waste as well

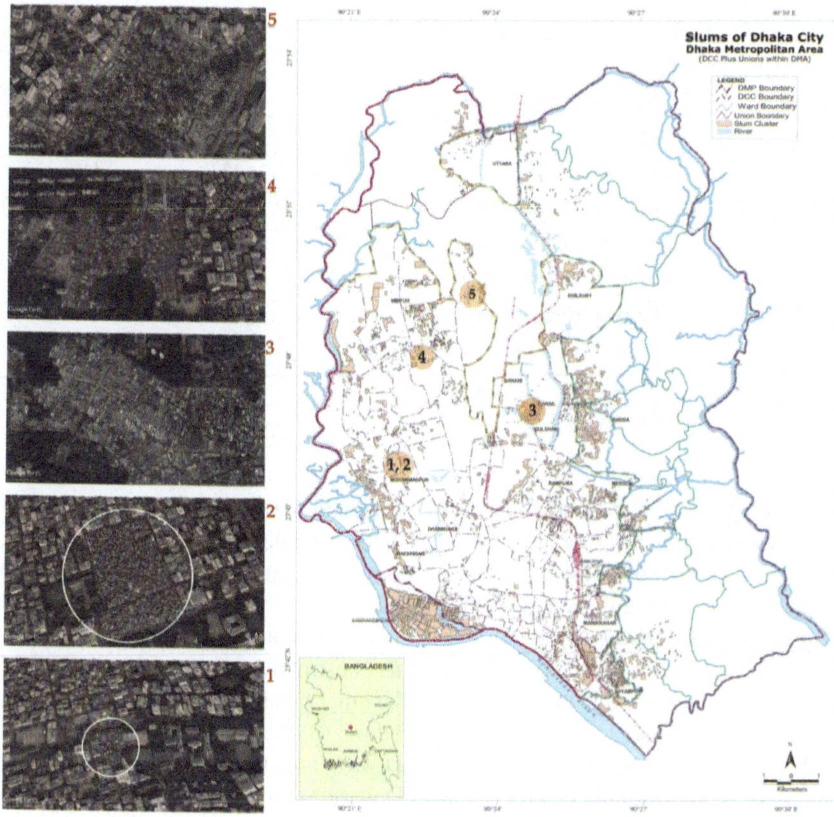

Fig. 1 Selected 5 sites in the map of slums of Dhaka City (CUS 2005); modified by Author

as modification of the dilapidated slum houses with recycled C&D waste in Dhaka simultaneously.

To collect data on knowledge, preference, and existing use of the C&D waste, 300 houses were surveyed in Mohammadpur Town Hall Camp (25 houses), Mohammadpur Geneva Camp (50 houses), Kallyanpur Slum (75 houses), Saat Tala Karail Slum (75 houses), and Bhashantek Slum (75 houses) (Table 2). Each slum area had a random sample of houses selected for the questionnaire survey depending on the availability and willingness of the slum dwellers to participate in the survey.

3.3 *Data Analysis*

Data from the questionnaires were compiled and analyzed in Statistical Package for the Social Sciences (SPSS, version 23) and Microsoft Excel. The correlations between the variables were analyzed using cross-tabulation with Cramer's V. Usually,

Table 2 Demographic profile of the respondents in the study areas

Demographic parameters		Number	%
Population	Number of respondents	300	–
Gender	Male	168	56%
	Female	132	44%
Income group	Low-income group (5–10 K)	80	27%
	Lower middle-income group (10–25 K)	205	68%
	Upper middle-income group (25–50 K)	15	5%
Age	Early working group (15–24 years)	87	29%
	Prime working group (25–54 years)	149	50%
	Mature working group (54–65 years)	51	17%
	Elderly (>65 years)	13	4%

Cramer's V 0.00–0.19 indicates a negligible correlation; Cramer's V 0.20–0.29 indicates a weak correlation; Cramer's V 0.30–0.49 indicates a moderate correlation; Cramer's V 0.50–0.69 indicates a strong correlation, and Cramer's V 0.70–1.00 indicates a very strong correlation (McHugh 2018). In this research, the statistical significance was defined as a p-value less than 0.05 (Fatemi et al. 2020). Respondents' "Preference for recycled C&D waste as construction products" was reported as the independent variable in the study. On the other hand, (1) the Age of the house, (2) the Construction material collection procedure, (3) the Familiarity with recycled C&D waste, (4) the Quality of recycled C&D waste, (4) the Cost of recycled C&D waste, and (5) the Durability of recycled C&D waste—serve as dependent variables.

4 Findings

4.1 *Policy Review on Circular Management of C&D Waste in Dhaka*

Initially, sixteen government policy instruments were identified related to waste management. However, nine regulatory documents out of these identified 16 documents deemed to have little relevance to C&D waste management were eventually eliminated after closer inspection. Lead Acid Battery Recycling Related Circular 2006, Medical Waste (Management and Handling) Rules 2008, Hazardous Waste and Ship Breaking Waste Management Rules 2011, Ship Breaking and Recycling Rules 2011, Electrical and Electronic Product Induced Waste (E-waste) Management Rules 2017, for instance, were specifically left out since they were deemed irrelevant to the study's topic and objectives. On the other hand, although two long-term national environmental policies were evaluated, the other four rules and acts pertaining to environmental management and environmental conservation were omitted, because these documents had been rendered updated by more contemporary waste management-specific rules and acts. Therefore, six documents, shown in Table 1, were considered for this study and subsequently evaluated since these fulfilled the review criteria. Moreover, Table 3 summarizes the findings from these policy analyses across four categories: construction products, C&D waste collection, C&D waste recycling, and alternatives to C&D waste. It can be seen from Table 3 that only the "New Clean Dhaka Master Plan 2018–2032" addressed three of the categories specified. It is the only document that addressed the category dealing with C&D waste alternatives, whereas the other documents addressed two categories each.

Clean Dhaka Master Plan 2005–2015 is one of these above-mentioned documents that was drafted to guide the City's waste management. As the first document of its sort, this document has its flaws, such as failing to address C&D waste explicitly and instead classifying it as either industrial waste (such as glass) or street garbage (such as sand, dust, etc.) (DCC 2005). However, plans for waste collection and recycling were discussed in detail, with diagrams and goals for the future included. There is no mention of possible substitutes for C&D waste in this document. The National 3R Strategy for Waste Management, implemented in 2010, was a radical policy instrument for Bangladesh because of the guiding principles and tactics it used to spread the 3R philosophy. It was indicated to improve informal recycling in slums and squatter communities by providing access to affordable and suitable technology (DoE 2010). It also recommended promoting research and development (R&D) by academies, research institutes, and even industrial institutes to achieve resource efficiency and, consequently, sustainable waste management. It further suggested that city corporations and municipalities of local governments should impose extra fees or administrative costs for the disposal of demolition debris to motivate homeowners and builders to produce less waste during the construction process [pg. 139] (DoE 2010). It did, however, exclude any discussion of C&D waste alternatives.

Both the 2013 and 2018 versions of the National Environmental Policy have one distinctive feature: they both advocate reducing the use of wood and brick in construction and increasing the production of alternatives of those building and construction products (e.g., Particle board, Block brick, Hollow brick) [pg. 10, 35] (DoE 2013), [pg. 69, 76] (DoE 2018). These documents also recommended the use of the 3R principles of reduce, reuse, and recycle in the housing, urbanization, and building industries [pg. 11] (DoE 2013), [pg. 22] (DoE 2018). Additionally, the National Environmental Policy 2018 urged that eco-friendly methods be included in urban slum redevelopment projects. However, neither version had a plan for how to collect C&D waste or how to deal with its alternatives.

Moreover, the New Clean Dhaka Master Plan 2018–2032 was designed and drafted with updated policy guidelines and cutting-edge techniques. It categorized the various forms of waste and assigned specific tasks to various organizations in charge of waste collection. For example, city corporations and municipalities of local government are responsible for overseeing the management of C&D waste, which is categorized as industrial waste collected during building construction (DNCC 2019). It further stated that only authorized entities may handle C&D waste in any

Table 3 Policy review on circular management of C&D waste in Dhaka

Policy	Relevant policy goals/strategies	Policy on non-renewable construction products' use	Policy on C&D waste collection	Policy on C&D waste recycling	Policy on alternatives of C&D waste
Clean Dhaka Master Plan 2005–2015	Develop the skillsets of Dhaka City Cooperation (DCC) employees and introduce cutting-edge technologies to create a comprehensive strategy for Dhaka City to manage solid waste by 2015	✗	✓	✓	✗
National 3R Strategy for Waste Management 2010	Foster a sustainable waste management system by addressing the challenges impeding 3R (Reduce, Reuse, Recycle) promotion in Bangladesh, defining the responsibilities of stakeholders in promoting 3R, and developing favorable environments for 3R implementation	✗	✓	✓	✗
National Environmental Policy 2013	Preserve natural resources through moderate urbanization and construction of additional new buildings, and ensure the effective use of resources, materials, and energy by implementing the 3R Principles: Reduce, Reuse, Recycling	✓	✗	✓	✗
National Environmental Policy 2018	Implement the policy of low consumption, reuse, and recycling in resource utilization, and encourage the sustainable use of all new and renewable resources	✓	✗	✓	✗

(continued)

Table 3 (continued)

Policy	Relevant policy goals/strategies	Policy on non-renewable construction products' use	Policy on C&D waste collection	Policy on C&D waste recycling	Policy on alternatives of C&D waste
New Clean Dhaka Master Plan 2018–2032	Mitigate environmental impact using advanced waste technology, establish an integrated and sustainable Solid Waste Management (SWM) framework for the city of Dhaka, and encourage citizen participation in SWM under the banner of "Zero-Waste"	✗	✓	✓	✓
Solid Waste Management Regulations 2021	Apply 6R Principles: Rethink, Refuse, Reduce, Reuse, Recycle, and Recovery for Solid Waste Management Instead of 3R (Reduce, Reuse, Recycling) in urban waste management to recover resources from solid waste	✗	✓	✓	✗

capacity, from collection to processing to final disposal. It plans to set up "construction waste recycling centers" with a capacity of 360 tons per day by 2032 when recycling is expected to have increased from its current 10 to 44%. More importantly, it detailed the components that may be recycled from C&D waste, such as recycled crushed stones, iron scraps, brick, sand, and soil [pg. 139] (DNCC 2019). Solid Waste Management Regulations 2021 recommended collecting and disposing of C&D waste separately from other waste [pg. 20495] (MEFCC 2021). It further suggested sorting and proper segregation before transferring it to the local government to prevent sand and dust particles to spread by wind or enter the sewers by rainfall. It, like the preceding documents, encouraged the adoption of the 3R principles of reduce, reuse, and recycling but skipped the policy on alternatives to C&D waste. The findings are summed up in Table 4.

Table 4 Summary of the policy review

CE strategy	Description of individual strategies	Policy
Policy on non-renewable construction products' use	– To reduce the use of wood and brick as raw materials in construction – To increase the production of alternatives of wood (e.g., Particle board) and brick (Block brick, Hollow brick)	National Environmental Policy 2013, National Environmental Policy 2018
Policy on C&D waste collection	– To collect and dispose of the C&D waste separately from other waste – To be handled by the authorized entities only in any capacity, from collection to processing to final disposal – To sort and segregate properly before transferring it to the local government agencies	Clean Dhaka Master Plan 2005–2015, National 3R Strategy for Waste Management 2010, New Clean Dhaka Master Plan 2018–2032, Solid Waste Management Regulations 2021
Policy on C&D waste recycling	– To adopt the 3R principles: reduce, reuse, and recycling in the housing development, urbanization, and construction industries – To improve informal recycling in slums and squatter communities by providing access to affordable and suitable technology – To set up C&D waste recycling centers to increase the current rate of recycling – To promote research and development (R&D) to achieve resource efficiency and consequently, sustainable waste management	Clean Dhaka Master Plan 2005–2015, National 3R Strategy for Waste Management 2010, National Environmental Policy 2013, National Environmental Policy 2018, New Clean Dhaka Master Plan 2018–2032, Solid Waste Management Regulations 2021
Policy on alternatives of C&D waste	– To encourage the use of recycled crushed stones, iron scraps, brick, sand, and soil as recyclables from C&D waste – To promote research and development (R&D) to explore more alternative uses for recycled C&D waste	New Clean Dhaka Master Plan 2018–2032

Table 5 Correlational Value of Attributes with Familiarity of Slum Dwellers

Attributes	Cramer's V	p-value	Total valid cases
Age of the house	0.360	0.000	300
Construction material collection procedure	0.441	0.000	300
Familiarity with recycled C&D waste	0.655	0.000	300
Quality of recycled C&D waste	0.454	0.000	300
Cost of recycled C&D waste	0.540	0.000	300
Durability of recycled C&D waste	0.391	0.000	300

4.2 *Preferences of Slum Dwellers to Use Recycled C&D Waste*

To understand the respondents' preferences for the use of recycled C&D waste, their responses were compared with six attributes: (1) the Age of the house, (2) the Construction material collection procedure, (3) Familiarity with recycled C&D waste, (4) the Quality of recycled C&D waste, (5) the Cost of recycled C&D waste, and (6) the Durability of recycled C&D waste. When analyzing the data using Cramer's V, a moderate to strong correlation has been found which is statistically significant (p-value < 0.05) (Table 5).

4.2.1 Respondents' Preference and the Age of Their Houses

The respondents' houses were categorized by their age using a ten-year interval, with the newest being less than ten years old and the oldest being more than fifty (Table 6). During the survey, it was revealed that the age of the house also determined how long a family lived in the same neighborhood. From frequency analysis, it was found that 0–9 years of houses 13%, 10–19 years of houses 16.7%, 20–29 years of houses 34.3%, 30–39 years of houses 13.3%, 40–49 years of houses 22.3%, 50+ years of houses 0.3%.

In cross-tabulation, a statistically significant correlation ($p = 0.000$) between the respondents' preferences to use recycled C&D waste and the age of their house has been found with a Cramer V of 0.360 (Table 5). Table 6 shows that people who live in relatively new houses are more enthusiastic about using recycled C&D waste than those who live in relatively older houses. Cumulatively, 64.1% of respondents (consecutively, 0–9 years 56.4%, 10–19 years 56.0%, 20–29 years 70.9%), whose houses built in the last 29 years, more likely to be interested in using recycled C&D waste, and cumulatively, 77.8% respondents (consecutively, 30–39 years 67.5%, 40–49 years 83.6%, 50+ years 100%), whose houses built between 30 and 50 years ago, more likely to be interested in using recycled C&D waste.

Table 6 Respondents' preferences and the age of their house

Preferences to use recycled C&D waste	Age of the houses													
	0–9 years		10–19 years		20–29 years		30–39 years		40–49 years		50 + years		Total	
	No	%	No	%	No	%	No	%	No	%	No	%	No	%
Yes	22	56.4	28	56.0	73	70.9	13	32.5	11	16.4	0	0	147	49.0
No	13	33.3	19	38.0	30	29.1	27	67.5	56	83.6	1	100	146	48.7
Refuse to respond	4	10.3	3	6.0	0	0	0	0	0	0	0	0	7	2.3
Total	39	100	50	100	103	100	40	100	67	100	1	100	300	100

4.2.2 Respondents' Preference and Construction Material Collection Procedure

Four types of collection procedures were identified based on material types and sources. The respondents: (1) bought new materials from the vendors, (2) collected new materials from different sources, (3) bought old materials from a recycling shop, and (4) collected old materials from the demolition site. From frequency analysis, it was found that the respondents bought new materials from the vendors 43%, collected new materials from different sources 27.7%, bought old materials from recycle shops 25%, and collected old materials from the demolition site 4.3%. In cross-tabulation, it was found that those who acquired old materials from a recycling shop and demolition site were more likely to be interested in using recycled C&D waste, whereas those who obtained new materials from vendors and different sources were less likely to be interested in doing so. This finding was statistically significant ($p = 0.000$), however, the correlations were moderate (Cramer's V of 0.441) (Table 5).

Table 7 shows that Around 93% of respondents (who bought old materials from recycle shops 91.4% and collected old materials from demolition sites 100%), who previously used old materials to build their houses, were enthusiastic to use those further. Even 35.4% of respondents (bought new materials from vendors 27.9% and collected new materials from different sources 45.0%), who used new materials to build their houses, during the survey, had shown their interest in using recycled C&D waste. However, 62% of respondents (who bought new materials from vendors 72.1% and collected new materials from different sources 49.0%), who previously used new materials to build their houses, were less interested in using recycled C&D waste.

Table 7 Respondents' Preference and Construction Material Collection Procedure

Preferences to use recycled C&D waste	Construction material collection procedure									
	New materials from vendors		New materials from different sources		Old materials from the recycling shop		Old materials from the demolition site		Total	
	No	%	No	%	No	%	No	%	No	%
Yes	36	27.9	45	45.0	**53**	**91.4**	13	100.0	147	49.0
No	**93**	**72.1**	49	49.0	4	6.9	0	0.0	146	48.7
Refuse to respond	0	0.0	6	6.0	1	1.7	0	0.0	7	2.3
Total	129	100	100	100	58	100	13	100	300	100

4.2.3 Respondents' Preference and Familiarity with Recycled C&D Waste

Familiarity with recycled C&D waste was measured in five categories: Extremely familiar, Moderately familiar, Somewhat familiar, Slightly familiar, and Unfamiliar. From frequency analysis, it was found that among the respondents 12% were extremely familiar, 26.3% were moderately familiar, 26.3% were Somewhat familiar, 30.7% were Slightly familiar, and 4.7% were unfamiliar with recycled C&D waste.

Based on the cross-tabulation analysis, a statistically significant ($p = 0.000$) strong correlation exists between respondents' preference and familiarity with recycled C&D waste (Cramer's $V = 0.655$) (Table 5). Respondents who were already aware of recycled C&D waste were more likely to express a preference for its usage than those who were not.

Table 8 reveals that 100% of respondents (100% of extremely familiar respondents and 100% of moderately familiar respondents with recycled C&D waste) were more interested in using recycled C&D waste. On the other hand, 91.5% (94.6% of respondents who were slightly familiar with and 71.4% who were unfamiliar with recycled C&D waste) were uninterested in using such materials (Table 8).

4.2.4 Respondents' Preference and Quality of Recycled C&D Waste

The quality of recycled C&D waste was evaluated in five categories: Very high, high, Average, Low, and Very low. From frequency analysis, it was found that the respondents felt 0.7% very high-quality, 22.7% high-quality, 39% average-quality, 34% low-quality, and 3.7% very low-quality recycled C&D waste.

According to cross-tabulation analysis, the association between respondents' preference and the quality of recycled C&D waste is statistically significant ($p = 0.000$) and Cramer's V ($V = 0.454$) indicates a moderate effect (Table 5). Table 9 demonstrates that respondents who preferred using recycled C&D waste had the experience of acquiring high-quality products previously. Of those surveyed, around 97.1% said they were interested in using C&D waste if they had access to high-quality products, whereas 81.4% of those who received low-quality items were averse to doing so (Table 9).

4.2.5 Respondents' Preference and Cost of Recycled C&D Waste

The cost of recycled C&D waste was further broken down into five categories: Very expensive, Expensive, Equal, Cheap, and Very cheap. From frequency analysis, it was found that the respondents felt 1% very expensive, 2.3% expensive, 25.3% equal, 69.3% cheap, and 2% very cheap recycled C&D waste.

The cross-tabulation analysis shows a statistically significant association between respondents' preferences and the cost of recycled C&D waste, $p = 0.000$. Cramer's V ($V = 0.540$) indicates that this association is strong (Table 5).

Table 8 Respondents' preference and familiarity with recycled C&D waste

Preferences to use recycled C&D waste	Familiarity with recycled C&D waste											
	Extremely familiar		Moderately familiar		Somewhat familiar		Slightly familiar		Unfamiliar		Total	
	No	%	No	%	No	%	No	%	No	%	No	%
Yes	36	100	**79**	**100**	30	38	2	2.2	0	0	147	49
No	0	0	0	0	49	62	**87**	**94.6**	10	71.4	146	48.7
Refuse to respond	0	0	0	0	0	0	3	3.4	4	28.6	7	2.3
Total	36	100	79	100	79	100	92	100	14	100	300	100

Table 9 Respondents' preference and quality of recycled C&D waste

Preferences to use recycled C&D waste	Quality of recycled C&D waste											
	Very high		High		Average		Low		Very low		Total	
	No	%	No	%	No	%	No	%	No	%	No	%
Yes	2	100	**66**	**97.1**	62	53	17	16.6	0	0	147	49
No	0	0	2	2.9	50	42.7	**83**	**81.4**	11	100	146	48.7
Refuse to respond	0	0	0	0	5	4.3	2	2	0	0	7	2.3
Total	2	100	68	100	117	100	100	100	13	100	300	100

Table 10 Respondents' preference and cost of recycled C&D waste

Preferences to use recycled C&D waste	Cost of recycled C&D waste											
	Very expensive		Expensive		Equal		Cheap		Very cheap		Total	
	No	%	No	%	No	%	No	%	No	%	No	%
Yes	0	0	**0**	**0**	2	2.6	**142**	**68.3**	3	50	147	49
No	3	100	7	100	**74**	**97.4**	62	29.8	0	0	146	48.7
Refuse to respond	0	0	0	0	0	0	4	1.9	3	50	7	2.3
Total	3	100	7	100	76	100	208	100	6	100	300	100

Table 10 indicates that if respondents discovered the cost of recycled C&D waste to be higher than the cost of equivalent new products, they did not use recycled C&D waste. Of those surveyed, around 97.7% of respondents (97.4% found equal, 100% found expensive and 100% found very expensive) found that the cost of recycled C&D waste is equivalent to or higher than the cost of equivalent new products (Table 10). The other 67.8% of respondents (68.3% found cheap, 50% found very cheap) found the cost lower than the cost of equivalent new products and therefore, wanted to use these further.

4.2.6 Respondents' Preference and Durability of Recycled C&D Waste

The durability of recycled C&D Waste was further divided into five categories: Very strong, Strong, Equal, Weak, and Very weak. From frequency analysis, it was found that the respondents felt 0.3% very strong, 0.3% strong, 52% equal, 46.4% weak, and 1% very weak recycled C&D waste.

The durability of recycled C&D waste was also statistically significantly associated with the respondents' preference, $p = 0.000$. According to Cramer's V ($V = 0.391$), the strength of this correlation is moderate (Table 5). Table 11 shows that respondents were interested in employing recycled C&D waste if they identified its

Table 11 Respondents' preference and durability of recycled C&D waste

Preferences to use recycled C&D waste	The durability of recycled C&D waste											
	Very strong		Strong		Equal		Weak		Very weak		Total	
	No	%	No	%	No	%	No	%	No	%	No	%
Yes	1	100	1	100	**116**	**74.4**	29	20.9	0	0	147	49
No	0	0	0	0	40	25.6	**103**	**74.1**	3	100	146	48.7
Refuse to respond	0	0	0	0	0	0	7	5	0	0	7	2.3
Total	1	100	1	100	156	100	139	100	3	100	300	100

durability to be comparable to that of similar new construction products. Nearly three-quarters (74.4%) of those surveyed felt that the durability of recycled C&D waste was on par with that of similar new building materials. Yet another 74.1% of respondents felt that recycled C&D waste was weaker than comparable new construction materials (Table 11).

5 Discussion

There has been a recent tendency in Dhaka City toward the construction of new high-rise buildings and the demolition of old and existing low-rise structures due to the city's high population density and rapid urbanization. Therefore, C&D waste has been produced as a byproduct and added to the waste stream and if not recycled, disposed of in landfills. Several waste management regulations of Bangladesh were analyzed to better comprehend the current state of C&D waste management and the potential for implementing the above-mentioned circular C&D waste management. The summary of the content analyses (Table 4) showed that the local government was aware of the need to collect, classify, and recycle C&D waste and that this effort would be multiplied by four by 2032. Recycled crushed stones, iron scraps, brick, sand, and soil are all examples of possible recyclables from these C&D waste that are presented in Table 4. Weil et al. (2006) corresponded that concrete and brick components can be recovered from C&D waste. Jamal and Sobuz (2017) reported that concrete and brick aggregates can be used for road base and pavement construction and iron scraps can be used at steel factories in Bangladesh. The summary also highlighted the need to consider wood and brick alternatives (Table 4). As brick alternatives, Paihte et al. (2019) proposed recycling plastic bottles filled with crushed aggregates recycled from C&D waste for building low-cost housing. Plastic lumber from recycled plastic bottles, coffee cups, etc. developed by Thomas Nosker may be used as a wood alternative and recycled again after they have served their original purpose (Dey and Iulo 2021). Finally, it is evident that the slum houses of Dhaka might benefit greatly from these recycled C&D wastes. Similar findings were reported by Elgizawy et al.

(2016), who suggested a "Learn to Earn Program" to redevelop slum houses and improve residents' economic standing by reusing C&D wastes.

For this research, it was also crucial to consider the slum residents' opinions on the utilization of recycled C&D waste in constructing and retrofitting their houses. Gálvez-Martos et al. (2018) concurred that the International Solid Waste Association lists "discussion, interaction, and user participation" as one of its good practice mechanisms for C&D waste management. Several attributes, including the Age of the houses, their Construction material collection procedure, their Familiarity with recycled C&D waste, the Quality of recycled C&D waste, the Cost of recycled C&D waste, and the Durability of recycled C&D waste, all played a role in shaping their preferences. The association of these attributes and residents' preferences was statistically significant ($p = 0.000$) with a moderate to strong correlation ($0.30 <$ Cramer's $V < 0.69$) (Table 5). From this empirical research, it has been established that 64.6% (12% + 26.3% + 26.3) of slum dwellers are familiar with recycled C&D waste, however, only 29.3% (25% + 4.3%) of them used it previously. The respondents' ensured that their preference highly depended on the quality, cost, and durability of the recycled C&D waste. As cited by Patel and Padhya (2021), if better quality materials can be generated through efficient recycling of C&D waste then that might be employed in the process of upgrading slum dwellings. Consistent with the previous findings, Thakuria et al. (2019) found that the government could upgrade more slum houses while reducing construction costs substantially if they used recycled C&D waste in the process. On the other hand, Bruen et al. (2013) reported that homeowners of low-income settlements in developing countries desired to construct a strong, contemporary house using durable materials like brick, cement and iron sheets despite the hefty cost of doing so. From these perspectives, this research may pave the way for future scholars, policymakers, as well as government and non-government organizations, to initiate the alternative application for the construction or renovation of slum houses, hence reducing C&D waste.

6 Concluding Remarks

This chapter set out to investigate a potential link between two crucial urban challenges in Dhaka, Bangladesh: Recycling huge quantities of C&D waste generated in Dhaka and upgrading the slum houses using this recycled C&D waste. The potential of recycling the C&D waste was gleaned through the analysis of several policy instruments of Bangladesh (Table 3). However, the establishment of construction waste recycling centers and ongoing R&D schemes, as outlined in the regulations (Table 4), will, nevertheless, expand its reach. Higher-grade construction products that can be developed from C&D waste need further study by the academic and research community. In addition, the scholars will have to come up with alternate building materials to limit the excessive usage of bricks and timbers. The slum residents interviewed in this research were quite enthusiastic to use recycled C&D waste as construction products. However, they desired to get the building materials to be cheaper and of better quality while still being durable or equivalent strength of similar new materials (Tables 9, 10, and 11). Access to affordable and suitable technology was highlighted in the regularity documents as a means to improve informal recycling in slums and squatter areas of Bangladesh (Table 4). If this informal sector is trained with affordable and suitable technology for recycling the C&D waste, there will be another two-fold benefit for the slum dwellers, as indicated in the "Learn to Earn Program" by Elgizawy et al. (2016): developing recycled materials (better to those currently in use) to upgrade their houses and generating income by selling the recycled materials (after fulfilling their need). In the end, this chapter allows for the development of a novel approach to the circular management of C&D waste and the enhancement of the efficiency of current resource management practices.

References

Ahmed HU (2019) Housing is a constitutional right. The Financial Express. Retrieved fromhttps://thefinancialexpress.com.bd/views/columns/housing-is-a-constitutional-right-1574259595

Ali ML, Akhter MW, Zobair MAA (2020) Housing for all: a better urban future. The Business Standard. Retrieved fromhttps://www.tbsnews.net/thoughts/housing-all-better-urban-future-151258

Amjad K (2021) Factors of climate migration of poor slum dwellers in Dhaka City. Int J Earth Sci Knowl Appl 3(3):273–288

Bruen J, Hadjri K, von Meding J (2013) Design drivers for affordable and sustainable housing in developing countries. J Civ Eng Arch 7(10):1220–1228

Chandran K (2016) Trash to treasure – C&D waste recycling story. Retrieved from https://www.devalt.org/newsletter/may16/of_5.htm

CUS (2005) Slums of urban Bangladesh: mapping and census. Centre for Urban Studies (CUS), Dhaka, Bangladesh

Datta SD, Rana MJ, Assafi MN, Mim NJ, Ahmed S (2022) Investigation on the generation of construction wastes in Bangladesh. Int J ConstrManag, 1–10. https://doi.org/10.1080/15623599.2022.2050977

DCC (2005) Clean Dhaka Master Plan 2005–2015 Dhaka, Bangladesh: Dhaka City Corporation (DCC), Government of the People's Republic of Bangladesh

Dey S, Iulo LD (2021) The circular economy of Dharavi: making building materials from waste. ARCC J Arch Res 18(2)

DNCC (2019) New clean Dhaka master plan 2018–2032. Dhaka North City Corporation (DNCC), Government of the People's Republic of Bangladesh, Dhaka, Bangladesh

DoE (2010) National 3R strategy for waste management. Department of Environment (DoE), Ministry of Environment and Forests, Government of the People's Republic of Bangladesh, Dhaka, Bangladesh

DoE (2013) National environmental policy 2013. Department of Environment (DoE), Ministry of Environment, Forest and Climate Change, Government of the People's Republic of Bangladesh, Dhaka, Bangladesh

DoE (2018) National environmental policy 2018. Department of Environment (DoE), Ministry of Environment, Forest and Climate Change, Government of the People's Republic of Bangladesh, Dhaka, Bangladesh

Dyer T (2012) Construction and demolition wastes. In: Meyers RA (ed) Encyclopedia of sustainability science and technology. Springer, New York, USA, pp 2391–2423

Elgizawy S, El-Haggar S, Nassar K (2016) Slum development using zero waste concepts: construction waste case study. Procedia Eng 145:1306–1313

EPA (2021) Sustainable management of construction and demolition materials. United States Environmental Protection Agency (EPA), Washington DC, USA. Retrieved from https://www.epa.gov/smm/sustainable-management-construction-and-demolition-materials

Fatemi MN (2012) Strategies to reduce construction and demolition (C&D) waste for sustainable building design in Dhaka: role of architects. Paper presented at the international seminar on architecture: education, practice and research, Dhaka, Bangladesh

Fatemi MN, Okyere SA, Diko SK, Kita M, Shimoda M, Matsubara S (2020) Physical vulnerability and local responses to flood damage in peri-urban areas of Dhaka Bangladesh. Sustainability 12(10):3957. https://doi.org/10.3390/su12103957

Favaretto P, Hidalgo GEN, Sampaio CH, Silva RdA, Lermen RT (2017) Characterization and use of construction and demolition waste from south of Brazil in the production of foamed concrete blocks. Appl Sci 7(10):1090. https://doi.org/10.3390/app7101090

Gálvez-Martos J-L, Styles D, Schoenberger H, Zeschmar-Lahl B (2018) Construction and demolition waste best management practice in Europe. Resour Conserv Recycl 136:166–178

Hasan MR, Sagar MSI, Ray BC (2022) Barriers to improving construction and demolition waste management in Bangladesh. Int J ConstrManag, 1–15. https://doi.org/10.1080/15623599.2022.2056804

He L, Yuan H (2020) Investigation of construction waste recycling decisions by considering consumers' quality perceptions. J Clean Prod 259:120928. https://doi.org/10.1016/j.jclepro.2020.120928

Inglis M (2007) Construction and demolition waste–best practice and cost saving. Paper presented at the regional sustainable building conference: sustainable building 2007 (SB07) Auckland, New Zealand

Islam R, Nazifa TH, Yuniarto A, Uddin AS, Salmiati S, Shahid S (2019) An empirical study of construction and demolition waste generation and implication of recycling. Waste Manage 95:10–21. https://doi.org/10.1016/j.wasman.2019.05.049

Islam S (2021) Urban waste management in Bangladesh: an overview with a focus on Dhaka. Paper presented at the 23rd ASEF Summer University (ASEFSU23) on 'liveable cities for a sustainable future', Singapore.

Jamal MS, Sobuz MHR (2017) A sustainable approach of demolition reconstruction, reuses and recycling construction materials in Bangladesh. Bangladesh Army Univ Eng Technol J 1(1):83–90

Jamil MA, Jamal MS (2018) An analysis for deconstruction and demolition of residential buildings in Dhaka City. Paper presented at the 4th international conference on advances in civil engineering 2018 (ICACE 2018), Chittagong, Bangladesh

Kibert CJ (2016) Sustainable construction: green building design and delivery. John Wiley & Sons, New York, USA

Kore SD, Vyas A (2016) Impact of marble waste as coarse aggregate on properties of lean cement concrete. Case Stud Constr Mater 4:85–92

Kulsum U, Moniruzzaman M (2022) Exploring the relationship of climate change and land-use dynamics with satellite-derived surface indices and temperature in greater Dhaka Bangladesh. J Earth Syst Sci 131(2):117

Liu J, Wu P, Jiang Y, Wang X (2021) Explore potential barriers of applying circular economy in construction and demolition waste recycling. J Clean Prod 326:129400

MCDC & ECD (2017) Waste management strategy and action plan for Mandalay. Mandalay, Myanmar: Mandalay City Development Committee (MCDC) and Environmental Conservation Department (ECD), Ministry of Natural Resources and Environmental Conservation (MONREC), the Republic of the Union of Myanmar

McHugh ML (2018) Cramér's V coefficient. In: Frey BB (ed) The SAGE Encyclopedia of educational research, measurement, and evaluation. SAGE Publications, New York, USA

MEFCC (2021) Solid waste management regulations 2021. Ministry of Environment, Forest and Climate Change, Government of the People's Republic of Bangladesh, Dhaka, Bangladesh
MoEJ (2019) Laws: waste & recycling law. Tokyo, Japan: Ministry of Environment of Japan (MoEJ) Retrieved from https://www.env.go.jp/en/laws/recycle/
Mulder E, de Jong TP, Feenstra L (2007) Closed cycle construction: an integrated process for the separation and reuse of C&D waste. Waste Manage 27(10):1408–1415
Nagapan S, Rahman IA, Asmi A, Memon AH, Latif I (2012) Issues on construction waste: the need for sustainable waste management. Paper presented at the 2012 IEEE Colloquium on Humanities, Science and Engineering (CHUSER), Sabah, Malaysia
Nipun MWH, Kakon AN, Hossain SN (2016) Growing residential built density of Dhaka city-causes, consequences and solutions: a case study on Mirpur section 2 and Kazipara. J Bangladesh Inst Plan 9:27–42
Ofori G, Briffett C IV, Gang G, Ranasinghe M (2000) Impact of ISO 14000 on construction enterprises in Singapore. Constr Manag Econ 18(8):935–947
Ouda O, Peterson H, Rehan M, Sadef Y, Alghazo J, Nizami A (2018) A case study of sustainable construction waste management in Saudi Arabia. Waste Biomass Valorization 9:2541–2555. https://doi.org/10.1007/s12649-017-0174-9
Paihte PL, Lalngaihawma AC, Saini G (2019) Recycled aggregate filled waste plastic bottles as a replacement of bricks. Mater Today: Proc 15:663–668
Patel Y, Padhya H (2021) Tactics of slum upgradation in urban area. Int J Res Eng Sci 9(2):48–53
PPCA (2018) Waste management strategy and action plan of Phnom Penh 2018–2035. Phnom Penh Capital Administration (PPCA), Phnom Penh, Cambodia
Purchase CK, Al Zulayq DM, O'Brien BT, Kowalewski MJ, Berenjian A, Tarighaleslami AH, Seifan M (2022) Circular economy of construction and demolition waste: a literature review on lessons, challenges, and benefits. Materials 15(1):76. https://doi.org/10.3390/ma15010076
Qinghai (2017) Turning waste into building materials: possible solutions for urban development. Hanoimoi
Rahman MM, Atkins PJ, McFarlane C (2014) Factors affecting slum sanitation projects in Dhaka city: learning from the dynamics of social-technological-governance systems. J Water, Sanit Hyg Dev 4(3):346–358. https://doi.org/10.2166/washdev.2014.081
Roychowdhury A, Somvanshi A, Verma A (2020) Another brick off the wall: improving construction and demolition waste management in Indian Cities. Centre for Science and Environment (CSE), New Delhi, India
Sanchez Rodriguez R, Ürge-Vorsatz D, Barau AS (2018) Sustainable development goals and climate change adaptation in cities. Nat Clim Chang 8(3):181–183
Sassi P (2019) Designing for dismantling, re-use and recycling. Greenspec
Shahidan S, Azmi MAM, Kupusamy K, Zuki SSM, Ali N (2017) Utilizing construction and demolition (C&D) waste as recycled aggregates (RA) in concrete. Procedia Eng 174:1028–1035
Shooshtarian S, Maqsood T, Caldera S, Ryley T (2022) Transformation towards a circular economy in the Australian construction and demolition waste management system. Sustain Prod Consum 30:89–106. https://doi.org/10.1016/j.spc.2021.11.032
Thakuria PP, Roy M, Dasgupta A (2019) Slum development using zero waste concepts. Paper presented at the 53rd international conference of the architectural science association, Uttarakhand, India
UN (2021) The government of Bangladesh announces its solid waste management rules 2021. Retrieved from https://sdgs.un.org/partnerships/government-bangladesh-announces-its-solid-waste-management-rules-2021
UNSD (2019) Report on sustainable development goal 11: sustainable cities and communities. Retrieved from New York, USA: https://unstats.un.org/sdgs/report/2019/goal-11/
WB (2022) Bangladesh development update: recovery and resilience amid global uncertainty. Retrieved from Dhaka, Bangladesh

WB (2023a) Industry (including construction), value added (annual % growth) - Bangladesh. Retrieved 4 April 2023, from World Bank https://data.worldbank.org/indicator/NV.IND.TOTL.KD.ZG?locations=BD

WB (2023b) Population living in slums (% of urban population). Retrieved 5 April 2023, from World Bank https://data.worldbank.org/indicator/EN.POP.SLUM.UR.ZS

Weil M, Jeske U, Schebek L (2006) Closed-loop recycling of construction and demolition waste in Germany in view of stricter environmental threshold values. Waste Manage Res 24(3):197–206

Whittaker MJ, Grigoriadis K, Soutsos M, Sha W, Klinge A, Paganoni S, …, Scullin M (2021) Novel construction and demolition waste (CDW) treatment and uses to maximize reuse and recycling. Adv Build Energy Res 15(2):253–269. https://doi.org/10.1080/17512549.2019.1702586

WPR (2023) World City Populations 2023. Retrieved 1 April 2023, from World Population Review (WPR)

Yang W-S, Park J-K, Park S-W, Seo Y-C (2015) Past, present and future of waste management in Korea. J Mater Cycles Waste Manage 17(2):207–217. https://doi.org/10.1007/s10163-014-0301-7

Zaman QM, Laing R (2013) Sustainable approach to regenerating residential form and density: case in Dhaka. Urban Stud Res

Md. Nawrose Fatemi completed his Ph.D. research on Climate Change and Densely Populated Urban Areas at Osaka University, Japan, as a MEXT (Monobukagoshu) Scholar. He has been serving the Department of Architecture, University of Asia Pacific (UAP), Dhaka since 2008 and presently holds the position as an Associate Professor. His research interest includes Urban Resilience, Sustainable Urban Design, and Climate Change and Vulnerability. He completed his master's degree in architecture (M. Arch) from the Bangladesh University of Engineering and Technology (BUET).

Tahmina Rahman is serving the Department of Architecture at the University of Asia Pacific as an Assistant Professor. She is an architect and urban designer by profession. She completed her master's research on Green-Blue Infrastructure of High-dense City at Osaka University, Japan. She has served several architectural organizations since 2008. She completed her undergraduate degree in Architecture (B. Arch) from the Bangladesh University of Engineering and Technology (BUET). Her research interest includes Green-Blue Infrastructure, Urban Regeneration, and Sustainable Built Environment.

Circularities in Housing Transformation Practices: A Synergetic Review

Florence Avogo Abugtane, Seth Opoku Mensah, and Gervase Kuuwaabong

Abstract Rapid urbanization, population growth, and a surge in household formation have outpaced the provision of affordable, decent housing for low-income city dwellers. These households have limited choices and transform existing structures to meet their housing needs. Though this housing transformation primarily focuses on meeting supply needs and strategizing adjustments, it also promotes the concept of circular and sustainable cities. Our review and analysis of the existing literature have allowed us to situate the principles of circularity within housing transformations, exploring the synergy between them. We found that households alter, extend, replace, and materially modify existing houses to meet housing needs. These practices align with the principle of repurposing. Furthermore, changes in construction materials contribute to enhanced durability and decreased reliance on finite natural resources, promoting sustainability. Moreover, this housing transformation process supports the reduction principle, as it fosters urban infilling, helping to prevent urban sprawl. When managed effectively, housing transformation can contribute to circular approaches that substantially reduce land and vegetation consumption, decrease waste, and aid in realizing circular and sustainable cities.

Keywords Circular cities · Housing transformation · Reduce · Repurpose · Synergic review approach

---

F. A. Abugtane (✉)
Department of Planning, SD Dombo University of Business and Integrated Development Studies, Wa, Ghana
e-mail: favogo@ubids.edu.gh

S. O. Mensah
Institute for Sustainable Futures, University of Technology Sydney, Ultimo, Australia
e-mail: seth.opokumensah@student.uts.edu.au

G. Kuuwaabong
Department of Geography, SD Dombo University of Business and Integrated Development Studies, Wa, Ghana
e-mail: gkuuwaabong@ubids.edu.gh

S. A. Okyere et al. (eds.), *Urban Slums and Circular Economy Synergies in the Global South*, Advances in 21st Century Human Settlements,
https://doi.org/10.1007/978-981-99-9025-2_6

1 Introduction

According to the United Nations Human Settlement Programme (UN-Habitat 2016), an alarming 980 million people lacked decent housing as of 2010, with a projected addition of 600 million people by 2030. This problem is compounded by unregulated rapid urbanization, which saw the number of slum dwellers in large and prime cities surge to 881 million between 2000 and 2014. (Permana and Wijaya 2017) suggest that this growth of urban populations has diverse impacts, including environmental degradation, cultural clashes, increased housing demand, and strained urban infrastructures. Providing adequate and affordable housing for low-income city residents remains a significant issue, as the United Nations Department of Economic and Social Affairs (United Nations Department of Economic and Social Affairs (UNDESA 2014) highlights. Pojani (2019) further clarifies that the urbanization process exacerbates the problem, with low-income city dwellers in need of housing often not provided by public initiatives. Market forces act as barriers, preventing these groups from accessing decent and affordable housing.

Private construction industries largely dominate the housing market in most countries. Yet, these industries tend to enable formal homeownership primarily for the middle class while often excluding the ever-growing population of low-income citizens from access to adequate and affordable housing (UN-Habitat 2016). This dichotomy fuels low-income city dwellers' adoption of various housing supply strategies. Pojani (2019) and Potts (2020) asserts that when low-income families are socially marginalized and unable to exercise their fundamental right to adequate housing fully, they resort to creating new urban occupations in informal settlements or modifying existing houses. Khan (2012) perceives this practice of creating additional space through housing extensions and alterations as a potential solution to reducing the housing supply deficit, especially in urban areas. It is, therefore, necessary to reduce the cost and increase the efficiency of housing production while also providing affordable housing alternatives for the financially disadvantaged. This can be accomplished through the effective and sustainable implementation of incremental housing and transformation practices, as outlined by (Bah et al. 2018).

Housing transformation involves physical alterations, extensions, and replacements (Tipple 2000). Vitasurya et al. (2018) defined these transformations as involving layout changes for improved functionality and the use of modern building materials, often without specific direction. (Portnov et al. 2005) stated that during these transformations, homeowners undertake maintenance and modifications of their apartments and houses to address various issues, from poor stock conditions to substandard housing design. Generally, the catalysts for housing transformation are diverse. They include the need to accommodate growing family sizes (Awanyo 2009; Isah et al. 2015; Piedra et al. 2020; Tipple 2000), economic motivations such as the desire for rental income or establishing home-based businesses (Asante and Ehwi 2022; Avogo et al. 2017; Kowaltowski et al. 2021), and socio-cultural reasons such as transforming spaces for worship (Alagbe and Aduwo 2014). Housing transformation also emerges as a response to the damage caused by natural disasters like floods (Wagemann 2017), for security reasons such as installing ironmongery (Adegun et al. 2019) and to adapt to a preference for a modern lifestyle (Khan 2012; Mirmoghtadaee 2009).

As the world continues to urbanize, there is a growing need to identify and implement innovative urban development models and strategies to address housing issues sustainably (Bîrgovan et al. 2022). In response to the multifaceted and urgent challenges posed by urbanization, (Lambropoulos 2021) and (Serpanos et al. 2022) suggest the concept of a circular city as an emerging solution in sustainable urban development. (Prendeville et al. 2018) define a circular city as one that applies principles of the circular economy to close resource loops in collaboration with key city stakeholders, thus paving the way for a resilient and sustainable future city. This circular approach can potentially address numerous socio-economic challenges plaguing cities, including providing affordable housing (Williams 2019). According to (Gravagnuolo et al. 2019), unused, abandoned, or degraded spaces can be valuable resources for urban regeneration and self-sustainability. These spaces can be creatively redesigned for new uses, injecting new life and potential value into cities and regions. In conclusion, (Kanters 2020) suggests that the principle of circularity in housing is centred on maximizing the use of available resources and minimizing waste and environmental impact in the construction process.

Existing literature on housing transformation primarily focuses on the factors driving transformation processes (Abed et al. 2023; Adegun et al. 2019; Asante and Ehwi 2022; Avogo et al. 2017; Khan 2014; Vitasurya et al. 2018), how occupants implement transformations through alterations, extensions, and replacements (Amiryar and Asano 2022; Kousa et al. 2021; Piedra et al. 2020; Tipple 2000), and the effects of these transformations on occupant well-being (Kowaltowski et al. 2021). However, previous studies have not directly addressed how housing transformation practices can contribute to the advancement of circular cities. To address the research gap, our chapter explores housing transformation practices and the principles of circular cities, ultimately illuminating the synergies between housing transformations and circular cities. This chapter expands our understanding of housing transformation beyond housing adjustments and supply strategies. It also has the potential to shape discussions and policies related to circularity and sustainability, particularly in rapidly urbanizing landscapes.

2 Methodology

We adopted a desk study approach to explore the synergies between the principles of circular and sustainable cities and housing transformation practices in developing countries. We examined these synergistic relationships through an extensive literature review covering the circular economy, circular cities, principles of circularity, and the operationalization and practices of housing transformation. We employed a combination of search methods, conducting electronic searches for relevant studies using Google Scholar, Scopus, and Web of Science, which are among the most current, comprehensive, and widely used search engines (Akinyi et al. 2021; Totin et al. 2018). In addition to this, we sourced pertinent studies by reviewing the reference sections of the collected papers. Our focus was on peer-reviewed articles, books, and book chapters published in English and focused on developing countries. To ensure we capture all relevant studies, we set our search period to include works published up to 2023. We excluded reviews, conference proceedings, working papers, editorials, and notes. We drew out synergies by comparing the identified studies with the indicators and principles of circular cities, seeking to understand how housing transformation practices align with these identified indicators. This process helped us understand the underlying concepts of the study, enabling us to identify the synergies between housing transformation practices and circular and sustainable cities.

3 Understanding the Concept and Principles of Circular Cities

Liu and Ramakrishna (2021) assert that a transition to a circular economy is indispensable for any economy striving to be competitive and develop sustainably while preserving our planet's life. The current linear take-use-dispose model is failing, making the shift towards a circular economy—where resources are retained within the system for more extended periods, and waste is minimized—an urgent necessity (Hoosain et al. 2021; Seetharam et al. 2021). Serpanos et al. (2022) highlight that the world's cities are experiencing exponential population growth, exacerbated by climate change and the relentless consumption of natural resources like water and land, leading to scarcity. Consequently, we must change how we build, develop, manage, and govern our cities, ensuring they are efficient, safe, and capable of meeting inhabitants' daily needs sustainably (Mersal 2016).

Achieving sustainable cities demands novel operational models encompassing the circular economy, resource management, and infrastructure development (Serpanos et al. 2022). Kisser and Wirth (2021) suggest that a circular city operates like an ecosystem, wherein waste always serves as a resource for something else. These cities thrive by preserving value and repeatedly generating value from the same material resources. Modak (2021) contends that the concept of a circular economy is multifaceted, converging management, resources, and residues in the service of the economy, livelihoods, and the environment. Thus, the pursuit of a circular economy is a vital strategy for achieving the Sustainable Development Goals (SDGs). Specifically, research by (Schroeder et al. 2019) indicates a positive and robust relationship between circular economy practices and the achievement of clean water and sanitation (SDG 6), as well as affordable and clean energy (SDG 7). The study further explains that by reusing and recycling resources, we can reduce the amount of waste and pollution, thereby positively impacting sustainable cities and communities and responsible consumption and production goals (SDGs 11 and 12). Additionally, by sparking new businesses and job opportunities, the circular economy can promote decent work and economic growth (SDG 8).

While the concept of a circular city stems from the circular economy model, where a boost in overall economic system productivity is anticipated by applying circular principles to current linear production-consumption patterns, its principles also apply to the spatial territorial dimension (Gravagnuolo et al. 2019). Despite being major resource consumers, cities serve as hubs for innovation with immense potential to lead the transition towards a circular economy and implement circular solutions such as creation, reuse, and recycling. The notion of cities becoming circular implies that city authorities encourage the reuse of building materials sourced from construction, renovation, and demolition projects. This, in turn, stimulates the market for recycled building materials (Bîrgovan et al. 2022).

Consequently, Kanters (2020) suggests that material selection should favour new bio-based materials from non-fossil sources, and materials derived from fossil fuels should be designed or reconfigured to enable their most effective reuse. Gravagnuolo et al. (2019) identify four specific strategies in the built environment sector to achieve circular cities: the circular design of new buildings, the dismantling of old buildings, the refurbishment of old buildings, and the reuse of old buildings. Consequently, circular cities feature modular and flexible buildings, designed for reuse or repurposing to ensure effective use, and disassembly rather than demolition to facilitate reuse and recycling (Byström 2018). Bîrgovan et al. (2022) underscore that the sustainable operation of smart technologies is an essential component of a circular economy and by extension, circular cities. Kisser and Wirth (2021) describe a circular city as a place where people share their resources, foster business models that prevent losses, and focus on maximizing resource productivity. Furthermore, Byström (2018) emphasizes that increasing circularity involves reusing idle or underutilized buildings and other assets within local value loops. Gravagnuolo et al. (2019) illustrates how rezoning a site previously used as a factory in the nineteenth century into a recreational space is an example of promoting reuse in circular cities. In another aspect, Hoosain et al. (2021) highlight that building materials possess a lifespan, termed the material circularity indicator, which can be determined through digital technology innovations. For instance, while concrete structures and brickwork are considered to have a lifespan of 100 years and are reusable after the building's use phase, aluminium and glass are expected to last 50 years.

Consequently, materials with longer lifespans are preferred for circular buildings as they are recyclable, eco-friendly, durable, and long-lasting (Hoosain et al. 2021). Furthermore, buildings, infrastructure, and products are designed for longevity in a circular city, implying that buildings should be durable and easy to maintain (Kisser and Wirth 2021). Governmental efforts towards achieving circular cities are evidenced in implementing policies like the home appliances old for a new initiative. When properly executed, this policy improves household appliances' energy efficiency and reduces environmental pollution (Yu et al. 2021). Practices such as repairing damaged devices to a usable condition, refurbishing, and remanufacturing also indicate the principle of circularity (Gheewala and Silalertruksa 2021; Modak 2021).

Cities, where land is scarce and finite, often prioritize high-value land uses; consequently, low-value activities that promote circular actions, such as urban farming, tend to lose out. As such, land is incorporated into the conceptualization of a circular approach to resource management (Williams 2019). Efforts to minimize land usage, such as installing solar panels on rooftops, thus eliminating the need for separate land for power plants, are part of the initiatives taken to promote circularity (Seetharam et al. 2021). Looping, another fundamental aspect of circularity, reduces resource wastage by closing resource loops through recycling, reuse, and energy recovery. Regeneration, on the other hand, involves the restoration of urban ecosystems, the preservation of natural capital, and the provision of vital ecosystem services. This can be achieved by integrating green and blue infrastructure, such as ponds and urban farming, into the urban fabric. Lastly, as a circular approach, adaptation entails planning and designing the city to enable the adaptation and renewal of existing infrastructure with minimal resource wastage. This often involves using flexible buildings to facilitate such adaptability (Williams 2019). These three actions—looping, regeneration, and adaptation—form the fundamental pillars of a circular approach to urban resource management.

In a broader context, the significance of incorporating circular economy practices into housing transformations is to adopt strategies aligned with circular building designs. Such strategies can substantially lessen the environmental impacts of buildings and reduce pressure on natural resources (Kanters 2020). As such, transitioning to a circular approach requires redesigning production and consumption systems, aiming to close the resource loop. This necessitates a multi-scale approach that extends from the household to the national level, involving policy formulations and ensuring active participation from multiple stakeholders (Modak 2021). Despite the importance of policy formation and stakeholder involvement, specific challenges may hinder the transition towards circularity at the household level and other sub-levels of engagement. These obstacles include difficulties in segregating waste, higher financial risks, limited availability of materials with high circular potential, policy, and technological constraints, and a lack of public participation (Kanters 2020; Schroeder et al. 2019; Sharma et al. 2021).

4 Housing Transformations and Practices

Housing transformation refers to homeowners undertaking construction activities to alter or extend their dwellings, effectively shifting from consumers to producers of housing. This process can be an efficient source of affordable housing, particularly for low-income individuals (UN-Habitat 2016). Housing transformations occur when additional spaces, rooms, or independent, functionally autonomous units are added to an existing house (Makachia 2006). The concept of housing transformation has been interpreted in various ways throughout literature. It has been defined as alterations, modifications, improvements, functional changes to rooms, and any alterations made to a house after habitation (Amiryar and Asano 2022; Tipple 2000). More specifically, it may involve adding new units or the subdivision of larger ones (Khan 2014). Makachia (2006) further refined this definition, identifying transformations as the lateral or vertical addition of spatial units like rooms, cubicles, and corridors. Additionally, transformations can involve the qualitative reorganization of existing spaces by relocating or resizing openings between spaces and/or to the exterior environment. Dweller-initiated transformations are notable for their positive contribution to housing stock and are increasingly seen as an alternative housing production model (ibid). Lastly, Amiryar and Asano (2022) concluded that low-income households frequently transform their homes through remodelling or extensions, tailoring their living spaces to suit their needs better.

Housing transformation has become indispensable in modern urban development (Rashid 2013). Strategies grounded in modernist templates, which overlook local cultural spatial paradigms, functional needs, and community consultation, are often faulted for the widespread prevalence of transformations (Abed et al. 2023; Makachia 2006). The failure to incorporate traditional lifestyles and socio-economic characteristics into housing designs has necessitated housing transformation (Amiryar and Asano 2022). Neglecting nonphysical aspects such as gender segregation, privacy, social networks, and safety drives the demand for layout transformations (Abed et al. 2023). Several factors influence housing transformations, including economic motives such as subletting for additional income and establishing home-based enterprises. Social motivations often relate to the need for accommodations that respect gender- and age-based cultural differentiation. On the other hand, physical factors generally relate to accommodating larger households (Khan 2014). Khan (2014) again highlights the role of ownership change, noting that new owners often seek transformations to suit their needs better. However, attaining quality norms in housing transformations often depends on income (Avogo et al. 2017). This suggests that economic factors also play a crucial role in determining the scale and quality of transformations in housing structures.

Housing transformation encompasses various practices driven by households' housing needs and contingent upon the financial capital at their disposal (Kousa et al. 2021). This concept entails extensions, alterations, replacements, and material changes to pre-existing housing structures. Amiryar and Asano (2022) posit that transformations develop over time and often involve more replacement than adaptation. For instance, detached houses might be demolished and substituted with apartment buildings. Kousa et al. (2021) further illuminate scenarios where houses suffering from varying degrees of destruction (from partial damage to complete ruin after a war) require replacements through transformation processes. Extensions, another common transformation practice, can manifest as vertical additions to existing structures. Informal annexes added for living or renting purposes also constitute a common form of transformation (Makachia 2011). Khan (2014) points out that rooftop additions, later converted into livable spaces or utilized for gardening, exemplify another transformation practice. According to Mirmoghtadaee (2009), a house during transformation resembles a living organism, growing and changing in response to new situations. These modifications often involve expanding or subdividing the existing structure to create new living spaces, ensuring the home continues to meet the evolving needs of its inhabitants.

Rashid (2013) and Amiryar and Asano (2022) highlight the critical role of construction materials in housing transformations. Originally, local materials such as jute sticks, bamboo, or mud were used for side walls, and hay or thatch served as roofing. However, these materials lacked durability. In the transformation process, concrete—made of cement, stones, and bricks—is now commonly used for walls, while corrugated iron sheets are used for roofing. As part of the transformation practice, these expensive, durable, and prestigious building materials have become the standard for shelter construction. Maintenance and modifications often involve substantial structural changes, such as large-scale demolition, rebuilding, and remodelling of properties. Common initiatives include replacing windows, renewing roofs, and installing central heating (Portnov et al. 2005).

Khan (2014) elaborates on the specifics of these transformations. Balconies, for example, are frequently converted into toilets, kitchens, or storage spaces. When extra space is available, it may be converted into small bedrooms or study rooms. Furthermore, these transformations often introduce amenities not initially present, such as in-house piped water and water-borne sanitation systems. These upgrades improve living conditions and add value to the property. While housing transformation practices are prevalent and crucial in meeting the housing needs of occupants, they are not without challenges. Several scholars have identified vital obstacles, including limited space availability, high cost of building materials, and enforcing building regulations and codes in certain jurisdictions. The ownership status of the house and the complex agreements required between vertical and lateral neighbours further complicate these transformations (Avogo et al. 2017; Makachia 2006, 2011; Tipple 2000).

5 Case Studies of Housing Transformation Practices

In a case study presented by (Makachia 2006), dweller-initiated housing transformations are highlighted as an alternative method of generating dwellings in cities like Nairobi, Kenya. This study explores transformations across four different types of housing. For single-story housing, transformations commonly involved the addition of extra spaces, rooms, or even independent, functionally autonomous units. Homeowners had the freedom to not only extend their own living spaces but also to create rental units for residential or other purposes. For multi-story housing, transformations typically involved vertical extensions, thus creating a spatial hierarchy. Public functions such as shops were often located on the street-facing ground-floor unit, while upper levels were primarily used for residential accommodation. This led to an unexpected functional mix. The study also observed that dwelling densities consistently increased with each addition in these multi-story transformations.

Additionally, the study noted that the building materials used in multi-story extensions were predominantly stable and permanent, such as steel-reinforced concrete and masonry. In courtyard houses, space sharing was a common feature. The courtyard acted as a communal area for laundry cleaning and drying, maintained through mutual arrangements. Consequently, transformations were typically avoided in these shared spaces.

The case study of housing transformation practices across different housing typologies in Nairobi suggests that several practices align with the principles of circular cities. For instance, vertical extensions in multi-story buildings, where ground-floors facing streets are used for commercial purposes and upper floors are reserved for residential uses, provide both accommodation and income-generating opportunities for residents. This mixed-use concept is crucial in current discussions about achieving circular cities. Transformations that increase residential densities in multi-story buildings align with densification strategies to address urban sprawl. Additionally, using durable, long-lasting materials such as concrete and steel in transformations and the shared use of courtyards minimizes the depletion of finite natural resources, including land. However, Makachia (2006) also noted potential challenges associated with unplanned transformations. These could compromise lighting and ventilation, leading to an over-reliance on artificial lighting. Despite these potential drawbacks, it can be generally concluded that housing transformation practices, as demonstrated in Nairobi, predominantly uphold the principles of circular cities.

In their recent research on housing transformation studies, Piedra et al. (2020) present a case study of the transformation of courtyard housing, known as "Cité," in Chile. The Cité consists of one or two-story houses, each with a built area of 50 to 80 m^2, arranged around a private alley that serves as the primary access. Typically, bedrooms, kitchens, and bathrooms are directly connected by an inner courtyard, which functions as a transition between the exterior and interior, facilitates illumination of adjacent areas, and aids natural ventilation. These houses, part of Santiago de Chile's architectural heritage, were built at the end of the nineteenth century. Over time, lack of maintenance and reduced housing space has led to the deterioration of these buildings. This has forced occupants to transform the existing courtyards to meet their needs.

Generally, the transformations in the Chilean case involved informal alterations, such as roofing the inner courtyards. However, these courtyards' reduced dimensions and modifications often led to insufficient natural light, leaving adjacent enclosures in near-complete darkness due to obstructed windows. This resulted in an over-reliance on artificial lighting and increased energy demand. Moreover, the modifications have led to airtightness in the homes, making them less energy efficient. However, the study noted that vegetation in these transformed courtyards helps balance temperatures, hinting at potential contributions to a circular economy. In summary, while housing transformations in Chile present challenges, such as over-reliance on artificial lighting, which does not align with circular city principles, vegetation in transformed courtyards suggests some benefits of a circular economy.

6 Synergies Between Housing Transformations and Circular and Sustainable Cities

Lambropoulos (2021) suggests that circular cities are not limited to closing material loops in city functions such as building, construction, energy production, waste processing, and spatial planning. It also involves actions that shape a sustainable built environment and influence residents' behaviour within cities. Their research Amiryar and Asano (2022) notes that activities like drilling holes, building dams, and constructing new houses can have positive and negative impacts on the environment. Hence, to mitigate the environmental drawbacks of constructing new houses, which contribute to urban sprawl and damage to ecosystems and vegetation, the principles of renovation, refurbishment, and reuse of existing houses and spaces should be promoted (Kanters 2020).

Abed et al. (2023) argue that remaining in the same house for an extended period while making modifications to meet evolving housing demands aligns with the discussions on circular cities. These diverse practices and perspectives reveal the synergy between housing transformation and circular city principles. Transformations made to existing houses to satisfy occupants' housing needs directly support the principles of repair, refurbishment, or redesign. This extends the house's usability, avoiding the need for new construction. By transforming existing houses, demand for new houses decreases, thereby reducing reliance on natural resources and energy needed for new constructions.

The synergy between material changes in housing transformation and circular city principles is underscored by a range of researchers, including (Lambropoulos 2021; Kisser and Wirth 2021; Hoosain et al. 2021; Kanters 2020; Rashid 2013; Amiryar and Asano 2022). Housing transformation often involves alterations to the building materials used in the house's construction. During these transformation processes, there's typically a shift from relying heavily on finite natural resources, such as timber, to more durable materials. This shift aligns with the circular city's principle of reducing pressure on natural resources (Kanters 2020). Furthermore, their material circularity indicators dictate the preference for long-lasting, easy-to-maintain materials. Hence, the shift towards more durable materials in housing transformation practices corresponds with the assertions by (Kisser and Wirth 2021) and (Hoosain et al. 2021). These authors argue that the material circularity indicators of concrete structures and brickwork, which are deemed to have a lifespan of 100 years and are reusable at the end of the building's use phase, promote sustainability. Choosing materials like concrete over earth or mud not only fosters recyclability and longevity but also supports eco-friendly practices. In effect, using such circular building materials, which possess longer lifespans, significantly mitigates the environmental impacts of construction.

Modak (2021), Gheewala and Silalertruksa (2021) and Portnov et al. (2005) highlight the influence of technology and electronic devices in housing transformations. The desire for modern conveniences and lifestyle enhancements drives changes in the usage of home appliances and features, such as the installation of air-conditioning units, electric or liquefied petroleum gas stoves, and water-borne toilet facilities. These changes hold significant implications for the principles of circular cities. For example, using water closet toilet facilities simplifies the extraction of by-products, which can serve as resources for biodigesters or be recycled into biogas. The transition to using gas stoves denotes a shift from reliance on fossil fuels for cooking, conserving natural resources, and minimizing air pollution. These transformation practices align with the views of (Liu and Ramakrishna 2021) and (Seetharam et al. 2021) that circularity is driven by emerging technologies.

These technologies facilitate universal access to electricity and contribute to reducing air pollution, a principal core to circularity. Williams (2019) further posits that employing energy-efficient infrastructure leads to a decrease in the consumption of finite fossil fuels. In addition to reducing reliance on fossil fuels through technological devices, occupants tend to gravitate towards acquiring energy-efficient electronics to minimize energy usage. Moreover, the practice of repairing faulty appliances extends their lifespan and prevents their premature disposal. This approach encourages the reuse of these appliances, which contributes substantially towards realizing circular cities.

Housing transformation practices, particularly those involving replacement and vertical extensions, are pivotal in advancing the principles of circular cities by reducing land usage. As indicated by (Amiryar and Asano 2022; Khan 2014; Kousa et al. 2021; Seetharam et al. 2021), common transformative actions that contribute to city circularity often involve vertical extensions to pre-existing structures. Such vertical additions in housing transformations significantly boost the number of households that can be accommodated on a smaller plot of land. Apartment buildings require limited space, helping to curb urban sprawl. Thus, efforts to minimize land use equate to conserving finite natural resources, a critical component of initiatives promoting circular and sustainable cities. In this context, vertical extensions as a housing transformation practice have direct implications for the assertions made by (Modak 2021; Seetharam et al. 2021). They emphasize that achieving circular cities requires measures like recycling and other actions to reduce land usage and conserve natural resources, such as virgin materials, to foster circularity.

Drawing from research by (Khan 2014, Lambropoulos 2021; Byström 2018; Mirmoghtadaee 2009), alterations in housing transformations align closely with the principles of reuse and repurposing, which are key to the concept of circular cities. Structures that have fallen into disuse are often transformed through alterations or replacements and are repurposed to prevent non-utilization. This practice of repurposing underused sections of a house for new functions, instead of building new structures, is a clear example of circular action. Reusing building materials is a common aspect of transformations associated with replacements. For instance, sand from demolished sections of a house is mixed with cement and reintroduced into the construction cycle to prevent waste. As (Gravagnuolo et al. 2019) point out, the circular design of new buildings, dismantling old buildings, refurbishing existing structures, and reusing old buildings are the four key strategies for achieving circular cities in the built environment sector.

Several studies, including those by (Lambropoulos 2021; Gravagnuolo et al. 2019; Avogo et al. 2017), discuss the need for adopting mixed-use strategies, which involve the co-location of compatible land uses. This principle is evident in the transformation of multi-story buildings, where ground-floor units facing primary or secondary roads often become shops. At the same time, the upper levels serve as residential accommodations (Makachia 2006). This results in an unexpected mix of functions, embodying the principles of reuse and repurposing and contributing to the reduction of urban land and space use. Moreover, some transformations are aimed at minimizing the usage of scarce resources such as land. For instance, some households opt for solar power for lighting, leading to the placement of solar panels on rooftops. This use of rooftops implies a reduction in land usage, which would otherwise be required for the installation of electricity high tension poles and transformers if traditional hydroelectric power were used.

Table 1 summarizes housing transformation practices and their synergies with some principles of circular cities. In housing transformations, occupants often alter the functional use of spaces. The implication for circular cities is reusing spaces that could otherwise have been abandoned, thereby reducing waste, and preserving the natural environment. Extensions, specifically vertical additions in housing transformation, encourage the reuse of durable materials and support densification and mixed uses. The impact of these vertical additions for circular cities is a reduction in land use for single-unit houses and separate uses, thereby significantly reducing the environmental impact of buildings and contributing to economic and ecological sustainability. Changes in building materials, from those based on natural resources to new bio-based materials with long lifespans, extend material lifespans and reduce over-dependence on fossil fuels. In replacement transformations, detached buildings are replaced with apartments that can accommodate more people in a smaller space. As with hydroelectric power, the preference for installing solar panels on roofs negates the need for additional land for erecting tension poles. The use of technological equipment, like energy-efficient cooking stoves, avoids the use of fossil fuels for cooking. Such technological gadgets also promote repairs and refurbishments, extending the product's lifespan and usefulness. Transformation practices in the form of alterations, extensions, changes in material uses, and replacements contribute positively to the principles of circular cities through reuse, repair, refurbishment, reduction, recycling, and redesign of existing housing.

While housing transformation practices contribute positively to enhancing the principles of circular cities, it is essential to note that some practices may impede these goals. As Makachia (2006) and Piedra et al. (2020) have pointed out, some transformations can lead to spaces initially providing natural light and ventilation being covered, thereby reducing lighting and air circulation. Such practices result in over-reliance on artificial lighting and increase energy consumption, challenging the principles of circular cities. Instead of closing loops by reducing energy use, these transformations may result in wasteful energy expenditure for lighting and ventilation that could be avoided with more thoughtful planning.

Table 1 Synergies between housing transformations and circular cities

Operationalization of housing transformation	Transformation practices	Reuse	Reduce	Repurpose	Sustainability
Alterations Morris and Winter 1975; Portnov et al. 2005; Seek 1983)	Changes in functions of units	Reuse of buildings	Preserve natural capital and waste	Mixed land uses	Economic and environment
Extensions/additions (Khan 2014; Morris and Winter 1975; Portnov et al. 2005; Seek 1983)	Vertical extensions of apartments Roof-top extensions for gardening	Building materials	Apartments/rooms accommodate more people and minimize the use of land	Mixed land uses/ gardening preserves natural capital and ecosystem	Social environment/ Economic
Material changes (Amiryar and Asano 2022; Khan 2014; Makachia 2006)	Durable ones Demand for newer housing consumptions Smart technologies	Repair and refurbish digital devices and equipment	Reduce natural resource consumption materials by using durable and long-lasting building materials		Economic, social, and environment
Replacements (Amiryar and Asano 2022)	Detached houses replaced with apartment buildings/frontage shops and accommodation mixed-uses		The use of solar panels installed on roofs and no separate land required for power plants reduce resource consumption and waste by using smart technologies, durable building materials	Change in functions of building	Environment, social, and economic

7 Conclusion

Given the rapid rate of urbanization and the ongoing utilization of finite natural resources such as land, vegetation, and water, it is essential to consider how we construct residential properties and manage cities sustainably. As observed in various studies, the path towards sustainability relies heavily on circular cities (Lambropoulos 2021; Williams 2023). Housing transformation processes, including alterations (reuse/repurpose), vertical extensions, and replacements, support infill development and curb urban sprawl, helping to minimize the use of finite resources. Additionally, changes in construction materials from temporary to long-lasting, durable options enhance the circular nature of cities. Given that cities are the largest consumers of natural resources, it is crucial to encourage housing transformation practices that reduce land use. The process of housing transformation, often seen in existing or even older homes, promotes the concept of circularity. The key objective is to redesign, repair, and refurbish for reuse instead of abandoning these structures.

In the realm of housing transformation, the flexibility of construction allows for easy adjustments and functional changes (Byström 2018). Policymakers focused on housing, circular, and sustainable cities should consider actions like alterations, vertical extensions, replacing detached buildings with high-rises, transitioning from natural resources as building materials, and incorporating technology and digital assets. These actions serve as circular approaches towards achieving sustainability. Recently, Williams (2023) underscored the need for cities to enhance their resilience by adopting new strategies that alleviate the pressure on scarce resources and prevent excessive waste. However, Kanters (2020) emphasizes that the correlation between housing transformation and the advancement of circular cities also depends on building ownership, which can influence the shift towards circular buildings. Lambropoulos (2021) suggests that cities need to be ecologically regenerative and adaptive to achieve circularity. Therefore, residents should be educated and advised on the choices to make during housing transformation processes that enhance circular cities. This includes reducing the use of materials and energy intensity, avoiding non-biodegradable products and materials in transformations, and adopting reuse, repair, and extraction strategies to promote recycling. To advance circular city practices, Schroeder et al. (2019) highlight the need for more emphasis on skilled training, capacity building, design development, and multi-stakeholder collaborations. These efforts should be concerted at the household, regional, and national levels.

This review has established a robust relationship between housing transformation processes and the principles of circular cities. Going forward, it would be beneficial to conduct empirical studies specifically designed to explore the strategies adopted by homebuilders in housing development aimed at enhancing circular and sustainable cities. Additionally, it would be worthwhile to investigate existing policies and strategies at both national and regional levels and the options available at the local level to promote and enhance circular cities in developing countries. Finally, further research is needed to examine the role of multi-stakeholder engagement in promoting circular cities.

References

Abed A, Obeidat B, Gharaibeh I (2023) The impact of socio-cultural factors on the transformation of house layout: a case of public housing—Zebdeh-Farkouh, in Jordan. J Asian Architect Build Eng 22(3):1195–1208. https://doi.org/10.1080/13467581.2022.2074021

Adegun OB, Adedeji JM, Adedeji YMD (2019) Whither the courtyards? Understanding disappearance and transformation of courtyards in residential buildings in Akure, Nigeria. IOP Conf Ser Mater Sci Eng 640(1):012010. https://doi.org/10.1088/1757-899X/640/1/012010

Akinyi DP, Ng'ang'a SK, Girvetz EH (2021) Trade-offs and synergies of climate change adaptation strategies among smallholder farmers in sub-Saharan Africa: a systematic review. Reg Sustain 2(2):130–143. https://doi.org/10.1016/j.regsus.2021.05.002

Alagbe O, Aduwo E (2014) The impact of housing transformation on residents' quality of life: a case study of low-income housing estate, Ipaja, Lagos. Covenant J Res Built Environ (CJRBE) 2(2)

Amiryar MR, Asano J (2022) A study on transformation of housing typology and its environmental and social effects on the living conditions of residents in planned residential neighborhoods of Kabul City. Urban Sci 6(3)

Asante LA, Ehwi RJ (2022) Housing transformation, rent gap and gentrification in Ghana's traditional houses: Insight from compound houses in Bantama, Kumasi. Housing Stud 37(4):578–604. https://doi.org/10.1080/02673037.2020.1823331

Avogo FA, Wedam EA, Opoku SM (2017) Housing transformation and livelihood outcomes in Accra, Ghana. Cities 68:92–103. https://doi.org/10.1016/j.cities.2017.05.009

Awanyo L (2009) Meeting housing-space demand through in situ housing adjustments in the Greater Accra Metropolitan Area, Ghana. Eviron Plann C Gov Policy 27(2):302–318. https://doi.org/10.1068/c0779b

Bah E-HM, Faye I, Geh ZF (2018) Housing market dynamics in Africa. Springer Nature. https://doi.org/10.1057/978-1-137-59792-2

Bîrgovan AL, Lakatos ES, Szilagyi A, Cioca LI, Pacurariu RL, Ciobanu G, Rada EC (2022) How should we measure? A review of circular cities indicators. Int J Environ Res Public Health 19(9). https://doi.org/10.3390/ijerph19095177

Byström J (2018) The 15 circular steps for cities. European Investment Bank. https://doi.org/10.2867/39283

Gheewala SH, Silalertruksa T (2021) Life cycle thinking in a circular economy. In: Liu L, Ramakrishna S (eds) An introduction to circular economy. Springer Singapore, pp 35–53. https://doi.org/10.1007/978-981-15-8510-4_3

Gravagnuolo A, Angrisano M, Fusco Girard L (2019) Circular economy strategies in eight historic port cities: criteria and indicators towards a circular city assessment framework. Sustainability 11(13)

Hoosain MS, Paul BS, Raza SM, Ramakrishna S (2021) Material passports and circular economy. In: Liu L, Ramakrishna (eds) An introduction to circular economy. Springer Singapore, pp 131–158. https://doi.org/10.1007/978-981-15-8510-4_8

Isah AD, Khan TH, Ahmad ASB (2015) Design implications: Impact of socio-physical setting on public housing transformation in Nigeria. J Manage Res 7(2). https://doi.org/10.5296/jmr.v7i2.6925

Kanters J (2020) Circular building design: an analysis of barriers and drivers for a circular building sector. Buildings 10(4)

Khan MH (2012) Governance during social transformations: challenges for Africa. New Polit Econ 17(5):667–675. https://doi.org/10.1080/13563467.2012.732276

Khan TH (2014) Houses in transformation: search for the implicit reasons. https://doi.org/10.1007/978-3-319-02672-5

Kisser J, Wirth M (2021) The fabrics of a circular city. In: Liu L, Ramakrishna S (eds) An introduction to circular economy. Springer Singapore, pp 55–75. https://doi.org/10.1007/978-981-15-8510-4_4

Kousa C, Pottgiesser U, Lubelli B (2021) Post-Syrian war residential heritage transformations in the old city of Aleppo: socio-cultural sustainability aspects. Sustainability 13(21)

Kowaltowski DCCK, Muianga EAD, Ruschel RC, da Silva VG, Granja AD, Moreira DC, da Silva RF, Moreira JA, De Vasconcellos LHSC (2021) Housing transformation in search of well-being ENCAC2021

Lambropoulos J (2021) Neighbourhoods of the circular city—how circular city strategies are implemented at the neighbourhood level. A qualitative analysis of circular city strategy implementation at the scale of housing and the neighbourhood Radboud University. Nijmegen

Liu L, Ramakrishna S (eds) (2021) An introduction to a circular economy. Springer Singapore

Makachia PA (2006) Influence of house form on dweller-initiated transformations in urban housing. Int J Hous Sci Appl 30–31(1):33–44

Makachia PA (2011) Evolution of urban housing strategies and dweller-initiated transformations in Nairobi. City, Cult Soc 2(4):219–234. https://doi.org/10.1016/j.ccs.2011.11.001

Mersal A (2016) Sustainable urban futures: Environmental planning for sustainable urban development. Procedia Environ Sci 34:49–61. https://doi.org/10.1016/j.proenv.2016.04.005

Mirmoghtadaee M (2009) Process of housing transformation in Iran. J Constr Develop Countries 14(1)

Modak P (2021) Circular economy practices in India. In: Liu L, Ramakrishna S (eds) An introduction to circular economy. Springer Singapore, pp 555–575. https://doi.org/10.1007/978-981-15-8510-4_27

Morris EW, Winter M (1975) A theory of family housing adjustment. J Marriage Fam 37(1):79–88. https://doi.org/10.2307/351032

Permana AY, Wijaya K (2017) Spatial change transformation of educational areas in Bandung. IOP Conf Ser Earth Environ Sci 99(1):012029. https://doi.org/10.1088/1755-1315/99/1/012029

Piedra K, Bustamante W, Schmitt C (2020) Living and comfort conditions in heritage housing: the inner courtyard as a uniting element in the cités of Santiago Poniente. IOP Conf Ser: Earth Environ Sci 503(1):012010. https://doi.org/10.1088/1755-1315/503/1/012010

Pojani D (2019) The self-built city: theorizing urban design of informal settlements. Archnet-IJAR: Int J Architect Res 13(2):294–313. https://doi.org/10.1108/ARCH-11-2018-0004

Portnov B, Odish Y, Fleishman L (2005) Factors affecting housing modifications and housing pricing: a case study of four residential neighborhoods in Haifa, Israel. J Real Estate Res 27(4):371–408. https://doi.org/10.1080/10835547.2005.12091164

Potts D (2020) Broken cities: inside the global housing crisis, 1st edn. Bloomsbury Publishing

Prendeville S, Cherim E, Bocken N (2018) Circular cities: mapping six cities in transition. Environ Innov Soc Transitions 26:171–194. https://doi.org/10.1016/j.eist.2017.03.002

Rashid M (2013) The role of cultural heritages on the physical transformation of rural housing. J Today s Ideas-Tomorrow s Technol 1(2):99–112. https://doi.org/10.15415/jotitt.2013.12007

Schroeder P, Anggraeni K, Weber U (2019) The relevance of circular economy practices to the sustainable development goals. J Indus Ecol 23(1):77–95. https://doi.org/10.1111/jiec.12732

Seek NH (1983) Adjusting housing consumption: improve or move. Urban Stud 20(4):455–469. https://doi.org/10.1080/00420988320080811

Seetharam DP, Khadilkar H, Ganu T (2021) Circular economy enabled by community microgrids. In: Liu L, Ramakrishna S (eds) An introduction to circular economy. Springer Singapore, pp 179–199. https://doi.org/10.1007/978-981-15-8510-4_10

Serpanos D, Muñoz L, Chatzigiannakis I (2022) Smart and circular cities. Computer 55(12):38–41. https://doi.org/10.1109/MC.2022.3208843

Sharma HB, Vanapalli KR, Samal B, Cheela VRS, Dubey BK, Bhattacharya J (2021) Circular economy approach in solid waste management system to achieve UN-SDGs: solutions for post-COVID recovery. Sci Total Environ 800:149605. https://doi.org/10.1016/j.scitotenv.2021.149605

Tipple G (2000) Extending themselves: user initiated transformations of government-built housing in developing countries, 1st edn. Liverpool University Press. http://www.jstor.org/stable/j.ctt5vjjq2

Totin E, Segnon AC, Schut M, Affognon H, Zougmoré RB, Rosenstock T, Thornton PK (2018) Institutional perspectives of climate-smart agriculture: a systematic literature review. Sustainability 10(6):1990

UN-Habitat (2016) Urbanization and development: emerging futures (World Cities Report Issue). UN-Habitat

Vitasurya VR, Hardiman G, Sari SR (2018) Transformation of traditional houses in the development of sustainable rural tourism, case study of Brayut Tourism Village in Yogyakarta. IOP Conf Ser: Earth Environ Sci 106(1):012060. https://doi.org/10.1088/1755-1315/106/1/012060

Wagemann E (2017) Need for adaptation: transformation of temporary houses. Disasters 41(4):828–851. https://doi.org/10.1111/disa.12228

Williams J (2019) Circular cities. Urban Stud 56(13):2746–2762. https://doi.org/10.1177/0042098018806133

Williams J (2023) Circular cities: planning for circular development in European cities. Eur Plan Stud 31(1):14–35. https://doi.org/10.1080/09654313.2022.2060707

Yu K, Zhang H, Liu Y (2021) Recycling of waste electric and electronic products in China. In: Liu L, Ramakrishna S (eds) An introduction to circular economy. Springer Singapore, pp 241–262. https://doi.org/10.1007/978-981-15-8510-4_13

Circular Economy in Africa's Informal Cities: A Review of Residents' Value Retention Practices and Their Implications for Participatory Urban Planning

Stephen Leonard Mensah, Louis Kusi Frimpong, Seth Asare Okyere, Shine Francis Gbedemah, and Matthew Abunyewah

Abstract Spatial informal urbanism practices towards value retention and circularity have received little attention in extant literature. Yet, informal settlements in African cities have played out as built spaces for necessity-driven value retention of materials, goods, and services, which potentiate the circularity of waste resources. This chapter highlights informal settlement practices that contribute to value retention and circularity of waste materials in selected African cities. This perspective departs from and contests dominant but empirically misleading notions of informal settlements as the "habitus of environmentally deteriorating practices" in northern scholarship. Data for the chapter is derived from secondary sources such as journal articles and grey literature. The findings of the chapter reveal that spatial informal urbanism practices oriented towards meeting everyday basic socioeconomic needs in the form of sorting out waste for recycling, reuse, and repurposing support circular principles and are critical to realizing circular goals in Africa's informal cities. The

S. L. Mensah (✉)
Department of City and Regional Planning, The University of Memphis, Memphis, TN, USA
e-mail: slmensah77@gmail.com

L. K. Frimpong · S. Francis Gbedemah
Department of Geography and Earth Sciences, University of Environment and Sustainable Development, Somanya, Ghana
e-mail: lkfrimpong@uesd.edu.gh

S. Francis Gbedemah
e-mail: sfgbedemah@uesd.edu.gh

S. A. Okyere
School of Landscape Architecture and Planning, The University of Arizona, Tucson, USA
e-mail: sethaokyere@arizona.edu

M. Abunyewah
The Australian Center for Resilience Implementation for Sustainable Communities, Charles Darwin University, Darwin, Australia
e-mail: matthew.abunyewah@cdu.edu.au

S. A. Okyere et al. (eds.), *Urban Slums and Circular Economy Synergies in the Global South*, Advances in 21st Century Human Settlements,
https://doi.org/10.1007/978-981-99-9025-2_7

chapter invites scholars and practitioners to recognize and support informal urbanism practices that contribute to the circular cities agenda while addressing any trade-offs and disservices that negatively affect the capacity of informal settlement dwellers to advance circular initiatives from below.

Keywords Informal practices · Circular economy · Value retention · Waste · African cities

1 Introduction

The future of humanity lies in urban centres; hence, cities are crucial spaces for the realization of numerous urban aspirations envisaged by authorities and communities (UN-Habitat 2022). As the world's urban population is expected to reach 68% by 2050, it has become imperative for cities to develop community-driven planning responses targeted at emerging and complex socioeconomic urban challenges such as climate change impacts and systemic socio-spatial inequalities. This expected increase is projected around 2.2 billion urban residents who are predicted to reside mostly in the Africa and Asia regions (UN-Habitat 2022). According to Guterres (2022) ensuring equitable, inclusive, and sustainable urban development requires that community resilience attributes are made integral components of city plans, especially in the production, distribution, and consumption of resources and materials. Informal settlements in African cities have emerged as dominant spaces for creating necessity-driven value retention materials, goods, and services, which potentiate the circularity of resources especially waste materials (Geyer 2022; Korsunova et al. 2022). Yet, the everyday livelihoods of informal residents are confronted with harsh spatial realities and socio-environmental and economic marginalization's (Quaye et al. 2022).

As urban population continues to rise, concerns about the indiscriminate use of resources especially in cities raise for environmental sustainability and equitable socioeconomic development (Marini 2021). In response, the World Commission on Environment and Development published the Brundtland report in 1987, which has become the global footprint on sustainable development. The report indicates that only a paltry 6% of materials on earth are recycled (Haas et al. 2015). To put the key tenets of the Brundtland report into practice through reducing natural resource extraction and minimizing waste generation, several governmental and non-governmental actors and institutions like the Ellen MacArthur Foundation have proposed the principles of a circular economy (CE) (Ellen MacArthur Foundation 2010). These principles and practices are undergirded by the study of real-world organisms through a nonlinear feedback-oriented mechanism (Ellen MacArthur Foundation 2013). According to Kirchherr et al. (2017), the CE concept is an alternative economic model, which replaces the end-of-life or throwaway culture through the principles and practice of reducing, reusing, recycling, and recovery of materials in the production, distribution, and consumption of goods.

Circularity practices especially in African cities are crucial towards the realization of sustainable urban development by improving environmental quality, economic prosperity, and social equity for the benefit of current and future generations (Denedo and Egbon 2021). In this view of (see Geissdoerfer et al. 2017; Mies and Gold 2021) advancing CE as a sustainability model provides opportunities for a regenerative system where waste, resource input, energy leakages, and emissions are minimized through the circularity of materials and resources. Reducing the wanton extraction, production, and disposal of resources and materials is critical in realizing Africa's development blueprints such as the Agenda 2063. The adoption and implementation of CE principles in African cities have been saddled with challenges such as inadequate funding, poor infrastructure, and weak legal and institutional frameworks (Tahulela and Ballard 2020). In the face of these challenges, some African countries such as Ghana, South Africa, Ivory Coast, and Nigeria have made significant strides towards adopting circular approaches both officially and unofficially (Africa Circular Economy Alliance 2021).

Studies documenting the role of informal urbanism and informal experimental initiatives in sustainable urban development in African cities (see Quaye et al. 2022; Azunre et al. 2021, 2022; Lyons 2016; Sietchiping et al. 2010; Okyere and Kita 2016; Okyere et al. 2021) have focused on their contributions to achieving the sustainability targets and the sustainable cities agenda. What remains unclear is the linkages that exist between experimental spatial informal urbanism practices and circular principles. Understanding this synergistic relationship engenders pathways for achieving value retention and a greener economy through urban circularity of materials and resources. Against this backdrop, this chapter reviews informal residents' value retention practices from selected African cities and draws implications for urban circularity. The contributions of the chapter are threefolds: first, it extends the discussion on the role of informal settlements in the sustainable cities agenda by extricating the circularity principles embedded in these practices. Second, the perspectives shared by the chapter depart from and contest dominant but empirically misleading notions of informal settlements as the "habitus of environmentally deteriorating practices" in northern scholarship. Lastly, it offers valuable lessons for urban practitioners in developing and implementing a CE model, including the development of sustainable circular plans and responses that are inclusive and address the needs of vulnerable populations in Africa's informal cities.

This section is followed by a theoretical review of sustainable urbanism and circular practices. Following next is the methodology and a presentation of selected cases of informal urbanism experimental practices drawn from different African cities. The methodology follows next and a discussion of the synergies that exist between the selected cases and the principles of CE in a way that germinates ideas for building an inclusive, resilient, and sustainable urban development. Lastly, the conclusion and policy implications of the findings are presented.

2 Sustainable Urbanism and Circular Practices in African Cities

Slums and informal settlements have become an integral spatial component of Africa's informal cities (Akuoko et al. 2021). They have become spatial manifestations of a long period of informal urbanization. Despite the efforts of city authorities, though some of them radical and perpetuate exclusion, Africa's informal cities have witnessed a surge in slums and informal settlements, where about 70% of urban residents find shelter (UN-Habitat 2015). Informal settlements are expected to increase parallel to Africa's rapid urbanization (UN-Habitat 2016). They are characterized by poor housing and environmental conditions amidst chronic socioeconomic marginalization (see Weimann & Oni 2019; Quaye et al. 2022). For instance, they lack access to safe drinking water, tenure security, and adequate living spaces (Quaye et al. 2022). Despite these challenges, residents living in informal settlements are famous for their creativity and ingenuity, which provide alternative approaches to address the ever-growing socio-environmental problems facing African cities. According to the International Labour Organization (2018), about 60% of the world's population make their living in informal settlements. Waste pickers form part of the top three largest worker groups (Dias and Samson 2016). The urban scholarship largely classifies informal settlements as origins of urban environmental problems, yet evidence shows that informal residents' practices are crucial in achieving climate and environmental sustainability (Cesaro et al. 2018). Highlighting these practices and their implications for urban circularity provides opportunities for tapping into grassroots knowledge for designing policies and responses towards building sustainable urban futures.

The CE concept, despite the enormous opportunities it offers, remains unclear how it provides access to opportunities and social equity for the informal sector (Murray et al. 2017; Millar et al. 2019). The reason is that formalization activities result in the unequal sharing of costs and benefits (Conke 2018). Scarcity of resources in emerging economies, which include those in African cities, pushes residents to engage in resource conservation practices that lead to value retention and circularity (Korsunova et al. 2022). The exhibition of these creative practices in informal communities is crucial for developing a sustainable CE model, where they become integral stakeholders in achieving urban circularity and development (Lee et al. 2021). Despite their socioeconomic contributions to address urban problems, the burgeoning literature has paid little attention to their creative processes, leaving little room for building local community resilience against externally imposed ideologies about the role of informal communities (Rigg and Oven 2015; Okyere et al. 2021). Indeed, striving for social justice and resilience of informal settlements demands that CE practices are given the necessary recognition and opportunities for integration into urban planning responses (Kirchherr 2021). Furthermore, it also requires a community-based planning approach where informal residents are involved in decision-making processes and are beneficiaries of the opportunities that emanate from circularity practices (Kirchherr 2021).

Practicing urban circularity presents numerous opportunities for employment creation, improving livelihoods, reducing waste, and achieving cleaner cities, among others. In fact, CE is a viable economic model for the production of sustainable goods and services through minimizing resource consumption and waste generation by engaging in the recovery of waste, water, and energy production (D'Adamo 2019). Despite the aforementioned, little has been done from policy and practice circles to integrate circular practices in the everyday socioeconomic lives of residents in some sub-Saharan African countries (Bertelsmann-Scott 2020). It is, therefore, imperative to government and non-governmental agencies to advocate for the implementation and utilization of the CE concepts to realize optimum socioeconomic development (Debrah et al. 2022). The CE model is a viable option for achieving environmental sustainability, improving the safety of raw materials, inspiring innovations, enhancing socioeconomic growth, and increasing the competiveness of industry players (Kalmykova et al. 2018; Elkins et al. 2019; Allen et al. 2019).

Efforts towards the full adoption and implementation of CE principles in African cities have been an uphill task due to a mirage of challenges. For instance, Ayeleru et al. (2020) and Diaz (2017) have identified insufficient funding, low capacity of personnel, lack of political will, and inadequate or absence of policies on circularity practices. In SSA, whereas countries such as Sudan, Sierra Leone, Guinea, and Somalia lack adequate national policies on circularity practices (Somé 2018), South Africa, Ghana, Ivory Coast, Nigeria, and Rwanda have formed an alliance that provides opportunities such as seminars, workshops, and training programmes to enhance circularity practices in member countries (Africa Circular Economy Alliance 2021). The alliance also engages in the creation of coalitions for the implementation of planned projects on CE, encouraging partnerships for achieving sustainability and facilitating the development of the next generation of African leaders who will champion the implementation of CE (Debrah et al. 2022).

Slums and informal settlements in Africa's informal cities have become exemplar urban spaces, which exhibit ingenuity and creativity through experimental informal resident's initiatives (Okyere et al. 2021; Okyere and Kita 2016). Such activities have contributed to waste reduction, climate change mitigation and adaptation, and building community resilience. Informal settlers continue to face systemic exclusions and marginalizations from city authorities as well as poor socio-environmental conditions. Yet, they provide opportunities for knowledge co-creation imminent for community-led planning responses for circularity transitions in Africa's informal cities.

3 Methodology

The main source of information used for this chapter was drawn from secondary literature such as journal articles and documents from international organizations such as the United Nations, the Africa Circular Economy Alliance, and the Ellen McArthur Foundation to produce a narrative review. The use of narrative review is crucial for conducting documentary analysis where it is difficult to have access to scholarly evidence (Okyere et al. 2023; Abunyewah et al. 2023). The process of the narrative review involved a careful search of peer-reviewed articles from the Scopus and Google Scholar databases. The Scopus database was used because it is reliable and is widely accepted in the academic community as a repository of peer-reviewed articles. Journal articles from Google Scholar were also used to complement those that were obtained from Scopus because Google Scholar holds a large volume of published works that go beyond those hosted in Scopus. Search through the articles was underpinned by the use of keywords such as "informal settlements", "circular economy", "circularity", "climate change", "adaptation", "mitigation", "disaster", "resilience" and others which were considered to be very related to the topic were checked for in the titles, abstracts and the contents of the downloaded relevant literature. After the search, the results were filtered to remove conference proceedings and papers that were not written in the English Language. This was followed by a thorough screening of the abstract of the remaining papers. The search results ended with retaining only papers that specifically discussed case studies of informal settler practices in African cities and those that were related to urban agriculture and green space development initiatives, urban housing initiatives, urban transportation initiatives, and urban waste management initiatives. The snowballing technique was also used to follow up on the references of some of the selected journal articles. This gave the opportunity to locate other related literature that provided significant insight in addressing the chapter objective.

The selected articles were analysed using the Microsoft Excel application. The cases were organized in the Excel application based on the main themes of the results, thus urban agriculture and green space initiatives, urban housing initiatives, urban transportation initiatives, and urban waste management initiatives. Two cases that fall under each of the four (4) aforementioned themes were selected as an exemplar of informal settler practices and to demonstrate the critical role played by residents of informal settlements towards sustainable urban development. The focus of the review was on case studies of informal settler practices that contribute to urban circularity. This contributed to the themes that have been presented in the section below. At the end of the analysis, eight (8) cases were presented from Uganda, Kenya, Sierra Leone, and Cameroon.

4 Informal Residents Initiatives

4.1 Urban Agriculture and Green Space Development Initiatives

- *Dandora Community Garden:* This subsection documents informal settlement practices on urban agriculture and green space development initiatives in selected informal settlements in some African informal cities, such as Nairobi in Kenya

and Kampala in Uganda. To begin with, the Dandora community environmental initiative is located in Dandora, a neglected slum community in Nairobi. The community is located closer to one of the city's biggest refuse dumpsites. Here, the residents have developed parks and gardens on the dumpsites by constructing drainage facilities and planting trees around them, hence rendering the barren community of Dandora a model of collective action (Tehlova 2019). This initiative was undertaken by the residents in order to change the negative perception of outsiders towards the community. They, therefore, achieve this by developing open spaces from recycled materials to protect the environment, encourage local entrepreneurship, and prevent the outbreak of diseases (Njeri 2020). The Dandora community environmental initiative has been significant for the following reasons; first, recycling has helped to minimize waste and greenhouse gas emissions as well as proven economic benefits (United Nations Economic and Social Commission for Asia and the Pacific 2017). Second, the initiative has also promoted innovative patterns of usage of waste materials from the dump site.
- *Innovative Vertical Farming:* This type of urban farming is practiced in the slums of Kampala City in Uganda. The farms are established from local materials such as plastics and wood (Wamboga-Mugirya 2019; Murungin 2021) and provide the following benefits to residents; thus, serve as a place for relaxation, generate income for the farmers and provide maximum utilization of urban space (Quaye et al. 2022). The farms also promote agro-tourism, which contributes crucially to local socioeconomic development, as well as strengthening social networking among the farmers (Quaye et al. 2022). Social capital among urban farmers, particularly those living in slums, improves and guarantees the resilience of informal urban households (Karunarathne and Lee 2022). In urban slums, where households are susceptible to varying socioeconomic, climatic, health, and environmental hazards, households' improved and assured resilience aids them to withstand any shocks or hazards that could strike on the community or the city.

5 Urban Housing Initiatives

- *Plastic Wall Housing in Kampala and Nairobi:* Plastics are substances that cause significant environmental problems due to their non-biodegradable nature. They are however important materials for housing construction in African slums (Pati et al. 2016; Akinwumi et al. 2019). The use of plastic bottles for building construction is considered cost-effective and more sustainable than the traditional bricks (Raut and Khan 2019). Findings by Kavarpuo (2022) in Bangladesh have corroborated the assertion that using plastics for housing construction is cost-efficient and durable. Residents in the slums of Kibeira have built walls made from plastic to replace houses built of mud, which were not resilient and durable (Kirui 2020). Using plastics to build walls for houses also contributes to reducing environmental pollution, as plastic pollution has become a major developmental issue in many African countries with many well-meaning citizens calling for a total ban on its use. Stressing on the use of plastics as a raw material for building construction, Cooke (2017) observes that it is used to build homes for many refugees in Algeria due to severe weather conditions. Furthermore, In Kibera, the use of plastic bottles has also been crucial in reducing environmental pollution (Quaye et al. 2022).

- *Earth bag Housing in Kenya:* Earth bags have become essential for housing development in the Nakuru slum of Kenya (Walker 2015). They are made from abandoned sacks collated from different parts of Kenya and filled with sand to create a room-like shape. Using earth bags is cost-efficient, beautiful, and environmentally friendly. It is also user-friendly and accessible (Quaye et al. 2022). Further, houses developed from earth bags are earthquake, fire, and climate resistant as well as thermally efficient as sand-filled earth bags are heavy and durable, which provides comfortable places of shelter for low-income households (Padgham et al. 2014; Belofsky 2018). Using earth bags for housing construction also expands the life span of houses as they take a long period to erode. While houses constructed from earth bags provide numerous benefits; it could be expensive when sacks are unavailable (Quaye et al. 2022). This could influence the ability of most slum households to put up affordable houses.

6 Urban Transportation Initiatives

- *Imagination car in Freetown:* The ingenuity and creativity of slum dwellers to provide services, which offer opportunities for their socioeconomic development is commendable. Residents living in the slums of Freetown have creatively made solar-powered mini cars from trash scraps and trash. The car runs 15 km/h and uses an engine powered by a roofed solar panel (Campbell 2021). This interesting innovation has also been replicated in the slums of other African cities (Jalloh 2020). Sierra Leone has poor transportation services and faces massive air pollution (World Health Organisation 2016). The solar-powered mini-cars are useful to slum residents in informal settlements because they can ply narrow roads and do not cause air pollution. It is also safe and conducive as it poses less risk for respiratory diseases (Quaye et al. 2022).
- *Makepes Eco-bins from Recycled Plastic Bottles in Cameroon:* Makepe is a slum community in Douala, Cameroon. Informal settlers have made a canoe known as ecobin from recycled plastic bottles. This innovation has become necessary to assist residents of the fishing community in transportation and minimize plastic pollution (Sawa 2014). The plastic materials are sourced from the drains, seashore, and the neighbourhood's environment by women, students, local fishermen, and residents (Goldapple 2020). The ecobins have become an important means of transportation on the water bodies in the community and are used to collect other plastic materials on the water. The ecobins have also become a much-needed innovation targeted at reducing plastic waste pollution and helping to protect water bodies. The canoes also create opportunities for exploring ecotourism in the community (Orayeva 2021; Moosmann 2021).

7 Urban Waste Management Initiatives

- *Garbage as Fuel in Nairobi:* Garbage generated in slum communities have become an important source of fuel for cooking in most slum communities. In Kibeira, it is a common practice to find residents in slum communities collecting waste from dumpsites to sell as cooking fuels to food vendors (Isaiah 2012). The cooking facility in the community has become a receptacle of various forms of waste, which is used as fuel for cooking. Residents send waste to the facility in exchange for cooking time tokens. For instance a 90-kg of waste is equivalent to an hour of cooking time (Sesay 2011; Quaye et al. 2022). This practice helps to reduce waste and the over-dependence on wood fuel for domestic activities such as cooking and heating (Quaye et al. 2022). It also prevents deforestation and promotes ecological sustainability. However, residents could be exposed to harmful and toxic chemicals such as highly combustible substances at the dumpsites or in the waste (Cesaro et al. 2018; Johnston et al. 2019).
- *Community Biodegradable Center in Uganda:* Residents of Kikooza slum located in central Uganda innovatively use biodegradable waste from various households to make animal feed. Household waste comprising of peels and food waste is dumped at a communal waste centre for further processing by residents engaged in animal husbandry (Mukama et al. 2016). Such a practice contributes to waste recycling and reuse. It also promotes urban agriculture, an important practice for building urban climate resilience in African cities (Ayambire et al. 2019; Azunre et al. 2021). Nearby shops in the community have also become important sources of organic waste to households, as slum residents gather organic waste from these shops with plastic bags. By so doing residents contribute to plastic waste reduction and promote responsible use of plastics (Mukama et al. 2016; Quaye et al. 2022). Such activities are crucial for minimizing environmental waste and as such complement city authorities' efforts in managing urban waste. It has also created employment opportunities for the majority of the youth who engage in waste sorting and picking of materials that can be recycled or used for composting. The use of garbage as cooking fuel instead of wood has also reduced dependence on wood, which promotes economic and ecological sustainability (Quaye et al. 2022).

8 Implications for Participatory Urban Planning

The chapter aimed to take stock of the case studies of value retention informal settlement practices in African cities and inherently draw their implications for urban circularity and sustainable urban development. The chapter has therefore shown that informal settlements, despite the widely held notion that are urban spaces where urban problems emanate have fascinating practices, which have the potential to address environmental, climate, and disaster concerns. In the subsequent paragraphs, these opportunities, which are inherently embedded in the case studies, are discussed. It also reveals the manifesting CE principles. This chapter submits and advocates for a new narrative that positions informal settlements and slums in Africa's informal cities as crucial urban partners for knowledge co-production towards building circular cities.

The findings from the stocktaking reveal that informal residents' activities are critical for promoting environmental sustainability. Rapid urbanization in Africa's informal cities have resulted in complex socioeconomic problems with devastating outcomes (Abunyewah et al. 2022a, b; Quaye et al. 2022). For instance, African cities have become hotspots for various forms of pollution (i.e., water, air, and environment) due to a linear system of waste and material disposal. The chapter findings show that the recycling of materials to protect the environment is an important practice that minimizes waste and leads to protecting open spaces, water bodies, and other forms of community resources (Njeri 2020). According to Njeri (2020), such activities targeted at using waste materials to protect open spaces and community gardening, prevent disease outbreaks. The review also shows that the Kampala innovative garden serves as a place of relaxation and agro-tourism centre for residents (Quaye et al. 2022). The presence of nature-based solutions in cities is crucial for sustaining their ecological integrity. In Kibeira, the use of garbage from dumpsites as fuel reduces reliance on wood fuel, prevents deforestation, and promotes the sustainability of urban ecological systems (Quaye et al. 2022). Such practice is imperative for sustaining ecological systems in African cities. Okyere et al. (2021) and Okyere and Kita (2016) in their study of ingenuities manifested by residents in an informal settlement in Ghana revealed similar practices. In view of the foregoing city authorities have the opportunity to design community-led planning approaches that position informal residents and communities as partners who possess important socio-ecological knowledge crucial for designing sustainable environmental transitions.

Residents in informal settlements have also become pivotal in minimizing environmental pollution through creative innovations. The Dandora community garden initiative proved to be an important innovation useful for minimizing waste generation and greenhouse gas emissions (United Nations Economic and Social Commission for Asia and the Pacific 2017). Activities that control air pollution in African cities need commendation as it is beneficial in reducing health risks and building sustainable cities and communities (see Fig. 1). For example, solar-powered cars developed by residents in Freetown minimize air pollution and lead to a low risk of contracting respiratory diseases (Quaye et al. 2022). Further, the use of plastics to design eco-bins by residents for transportation purposes has led to a drastic reduction in plastic pollution subsequently protecting the water bodies and creating opportunities for eco-tourism development (Orayeva 2021; Moosmann 2021). Reasoning with Cesaro et al. (2018), this finding validates the observation that informal settler activities are crucial pathways for achieving climate and environmental sustainability.

The review also shows that activities of informal settlers enhances the resilience of local communities against the impacts of climatic hazards and disasters. According to Walker (2015), the use of earth bags for building construction provides protection for

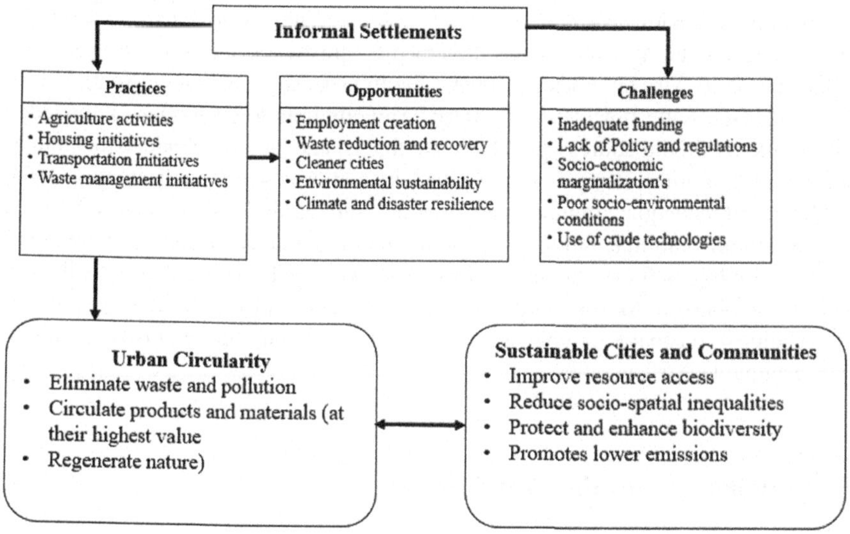

Fig. 1 Informal Settlements and urban circularity synergies (*Source* Author's elaboration)

local informal communities against the occurrence of climatic hazards and disasters. While it was found in Kibeira's informal settlements in Nairobi that the use of plastics for housing construction leads to weak capacity against the occurrence of climatic hazards and disasters, it was an important practice that results in the reduction of environmental pollution (Kirui 2020). A finding corroborated by Cesaro et al. (2018) and positions informal settlements as urban actors of change rather than despair. Informal settlements-led activities such as those documented by (see Walker 2015; Kirui 2020) have proven to be imperative bottom-up tactical approaches which inure to building climate and disaster resilience in local communities and among vulnerable populations in African cities (Hambati and Yengoh 2018; Satterthwaite et al. 2020). Considering the grave opportunities that emanate from informal residents' activities, it is imperative for city authorities in Africa to develop strategies, which reduce the socio-spatial inequalities and chronic economic marginalization characterizing informal settlements, and view them as crucial partners for building inclusive and sustainable cities.

Despite the crucial contributions of informal resident's activities in African cities as demonstrated by this chapter, city authorities and policymakers have paid little attention to them. The review has shown that informal urban spaces possess critical lessons useful for co-producing knowledge rich in community-led experiences and inputs. The chapter therefore challenges the traditional perception of informal communities as problem-originative areas and advocates a bottom-up approach in designing planning and policy responses for revitalization. By so doing, residents not only become partners for achieving sustainable urban development but provide a collaborative platform for addressing the socio-spatial inequalities and the economic marginalization they endure.

So far, the review has demonstrated viable circularity lessons embedded in informal resident's activities, even though they are by far the most marginalized urban spaces in African cities obviously due to aspects of their socio-spatial characteristics (see Satterthwaite et al. 2020). Any attempts to co-produce knowledge with residents should address any challenge that can act as a barricade to this much-needed collaboration. Here, we suggest that city authorities pay much attention to the challenges of inadequate funding, lack of policy and regulations on urban circularity, socioeconomic marginalization, poor socio-environmental conditions, and the use of crude technologies for waste and materials recoveries. Targeting these, challenges provide the platform for fostering informal settlements—city authorities' collaborative relationship capable of engendering ideas towards building inclusive, resilient, and sustainable African cities.

9 Conclusion and the Way Forward

In African cities, informal settlements and slums face numerous problems due to their legitimacy status, the challenges that characterize them, and the lack of recognition by city authorities as urban spaces that warrant planning responses. In spite of these, their everyday activities geared towards improving their socioeconomic activities have shown to possess the circular principles of recycling, reusing, rethinking, regenerating, and recovering. Indeed, it is not far-fetched to say there is a strong positive correlation between the activities of informal settlers and the principles of CE. This chapter aimed to review existing informal settler practices in African cities and draw their implications for urban circularity, a conduit imperative to achieving sustainable urban development. The findings of the study revealed that various initiatives comprising gardening and farming in informal settlements lead to waste reduction as waste products are used to create protection along water resources. It was also revealed that the development of such gardens and farms creates opportunities for the development of agro- and eco-tourism in these communities and leads to ecological sustainability. The chapter findings also show that informal activities lead to a reduction in pollution. Creative innovations such as the design and use of solar-powered vehicles and eco-bins for transportation purposes lead to a reduction in greenhouse gases and reduce the risk for respiratory diseases. The review also shows that creative ways of housing construction using resilient materials in informal settlements result in building their capacities against the occurrence of climate-related hazards and disasters.

The chapter based on the foregoing findings offers two major policy repercussions. First, informal community groups and actors, which comprise residents and opinion leaders, should be made an integral part of the urban planning process. In a consultative and collaborative orientation, they should be welcomed in making inputs into the design and implementation of city development plans. This offers opportunities for co-producing knowledge for sustainable urban development and is key to reducing the socio-spatial marginalization they endure. Lastly, government and non-governmental organizations should assist them with funding and logistical needs to expand their activities. Informal residents already have the experience and the setup to expand and create opportunities for many people. For example, receiving external support will create employment for the teaming urban youth, result in ecological protection and preservation, reduce environmental pollution, and build the resilience of informal communities against climate-related hazards and disasters.

Despite the significant contribution made by the chapter, it has two major limitations. First, it is based on a review of secondary literature and an inductive inference of the implications of informal sector practices for urban circularity and sustainable urban development. Future studies could therefore consider an empirical study of the linkages between informal practices and circularity practices. Second, it draws on selected case studies, which could limit its applicability. However, the lessons espoused in this chapter are useful for many SSA countries with similar characteristics.

References

Abunyewah M, Odei MEK, Okyere SA, Boateng FG (2023) Advancing slum-circular model for sustainability transitions in global south cities. Nat Sustain

Abunyewah M, Okyere SA, Frimpong LK, Diko SK, Erdiaw-Kwasie MO, Boateng V (2022a) Fire risk communication in the urban informal sector: evidence from traditional marketplaces in Accra. Risk, Hazards & Crisis in Public Policy, Ghana

Abunyewah M, Okyere SA, Diko SK, Kita M, Erdiaw-Kwasie MO, Gajendran T (2022b) Flooding in informal communities: residents' response strategies to flooding and their sustainability implications in old Fadama, Accra. In: Eslamian S, Eslamian F (eds) Disaster risk reduction for resilience. Springer, Cham. https://doi.org/10.1007/978-3-030-72196-1_18

Akinwumi II, Domo-Spiff AH, Salami A (2019) Marine plastic pollution and affordable housing challenge: shredded waste plastic stabilized soil for producing compressed earth bricks. Case Stud Constr Mater. https://doi.org/10.1016/j.cscm.2019.e00241

Africa Circular Economy Alliance (ACEA) (2021) Five big bets for the circular economy in Africa African circular economy alliance. https://www.afdb.org/en/topics-and-sectors/topics/circular-economy/african-circular-economyalliance-acea. Accessed 14 May 2022

Akuoko PB, Aggrey V, Amoako-Arhen A (2021) Ghana's informal economic sector in the face of a pandemic. Soc Sci Humanit Open, 3, Article 100094. https://doi.org/10.1016/j.ssaho.2020.100094

Allen D, Halloran P, Harris P, Kaduck J, Leith A, Lindsay C, Naylor W, Palmer S, Sismour K (2020) Acknowledgements the 2020 Vision Workgroup Mark McDermid, Wisconsin Department of Natural Resources. https://www.epa.gov/sites/default/files/2015 08/documents/sustainable_materials_management_the_road_ahead.pdf. Accessed 27 May 2022

Ayeleru OO, Dlova S, Akinribide OJ, Ntuli F, Kupolati WK, Marina PF, Blencowe A, Olubambi PA (2020) Challenges of plastic waste generation and management in sub-Saharan Africa: a review. Waste Manag 110:24–42

Azunre GA, Amponsah O, Takyi SA, Mensah H (2021) Informality-sustainable city nexus: the place of informality in advancing sustainable Ghanaian cities. Sustain Cities Soc 67, Article 102707. https://doi.org/10.1016/j.scs.2021.102707

Azunre GA, Amponsah O, Takyi SA, Mensah H, Braimah I (2022) Urban informalities in sub-Saharan Africa (SSA): A solution for or barrier against sustainable city development. World Dev, 152, Article 105782. https://doi.org/10.1016/J.WORLDDEV.2021.105782

Bertelsmann-Scott T (2020) Policy briefing the circular economy: including Africa in Europe' s circle. https://www.africaportal.org/publications/circular-economy-including-africa-europes-circle/. Accessed 20 May 2022

Campbell M (2021) First solar-powered car made out of trash invented in Sierra Leone [WWW Document]. https://www.euronews.com/green/2021/01/29/first-solar-powered-car-made-out-of-trash-invented-in-sierra-leone. (Accessed 10 March 2021).

Cesaro JD, Porphyre V, Duteurtre G (2018) Influence of the industrialization of pig farming in Vietnam on the diversification of systems in agriculture-livestock integration. J. Husb. Vet. Med. Trop. Ctries. 71(1–2):7–13

Cooke L (2017) Extraordinary man builds 25 plastic bottle homes for refugees in Algeria [WWW Document]

Conke LS (2018) Barriers to waste recycling development: evidence from Brazil. Resour Conserv Recycl 134:129–135. https://doi.org/10.1016/j.resconrec.2018.03.007

D'Adamo I (2019) Adopting a circular economy: current practices and future perspectives. Soc Sci 8:328

Debrah JK, Teye GK, Dinis MAP (2022) Barriers and challenges to waste management hindering the circular economy in Sub-Saharan Africa. Urban Science 6(3):57

Denedo M, Egbon O (2021) Africa, from the past to the present: Moving the critical environmental accounting research on Africa forward. In: Bebbington J, Larrinaga C, O'Dwyer B, Thomson I (eds) Routledge handbook of environmental accounting. Routledge, London

Dias S, Samson M (2016) Informal economy monitoring study sector report. Waste Pickers. WIEGO, Cambridge, MA, USA

Diaz LF (2017) Waste management in developing countries and the circular economy. Waste Manag. Res. 35:1–2

Ekins P, Domenech T, Drummond P, Bleischwitz R, Hughes N, Lotti L (2019) Managing environmental and energy transitions for regions and cities how and where background information managing environmental and energy transitions for regions and cities the OECD centre for entrepreneurship, SMEs, regions and cities on Twitter: @OECD_local Citation. http://t4.oecd.org/cfe/regionaldevelopment/Ekins-2019-Circular-Economy-What-Why-How-Where.pdf. Accessed 30 May 2022

Ellen MacArthur Foundation (2010) What is circular economy. Retrieved from https://www.ellenmacarthurfoundation/what-is the circular-economy, on 10.03.2023. org/circular-economy/

Ellen MacArthur Foundation (2013) Towards the circular economy. Opportunities for the competitive Europe. Ellen MacArthur Foundation/SUN/McKinsey Centre for Business and Environment. Accessed 20 March 2023, https://www.ellenmacarthurfoundation.org/assets/downloads/publications/EllenMacArthurFoundation_Growth-Within_July15.pdf

Ellen MacArthur Foundation (2015) Growth within: a circular economy vision for a consumer goods sector. Isle of Wight, UK. Ellen MacArthur Foundation

Geissdoerfer M, Savaget P, Bocken NMP, Hultink EJ (2017) The circular economy—a new sustainability paradigm? J Clean Prod 143:757–768

Geyer HS (2022) Can informality help create smart, sustainable cities? The vibrancy of self-organised informal settlements in Cape Town. Geo J, 1–19

Ghosh SK (ed) (2019) Sustainable waste management: policies and case studies: 7th IconSWM—ISWMAW 2017, Volume 1. Springer. https://doi.org/10.1007/978-981-13-7071-7_11

Guterres A (2022) Forword: world cities report 2022: envisaging the future of cities. Nairobi, Kenya, United Nations Human Settlements Programme (UN-Habitat)/United Nations Department of Economic and Social Affairs, Population Division

Goldapple L (2020) Cameroon's protector of the oceans: futureHero Ismaël Ebone. Atlas of the future. https://atlasofthefuture.org/cameroons-protector-of-the-oceans-futurehero-ismael-essome-ebone/

Hambati H, Yengoh GT (2018) Community resilience to natural disasters in the informal settlements in Mwanza City, Tanzania. J Environ Planning Manage 61(10):1758–1788

Haas W, Krausmann F, Wiedenhofer D, Heinz M (2015) How circular is the global economy?: an assessment of material flows, waste production, and recycling in the European Union and the world in 2005. J Industrial Ecol 19(5):765–777. https://doi.org/10.1111/jiec.12244

International Labour Organization (2018) Women and men in the informal economy: a statistical picture (third edition), Geneva

Isaiah E (2012) Garbage-fed community cooker cuts wood use, energy costs. United Nations Environment Programme. https://news.trust.org/item/20120410154500-sjgba. Accessed 15 Mar 2021

Jalloh AB (2020) Sierra Leone first solar car, Built from Trash [WWW Document]

Johnston HJ, Mueller W, Steinle S, Vardoulakis S, Tantrakarnapa K, Loh M, Cherrie JW (2019) How harmful is particulate matter emitted from biomass burning? A Thailand perspective. Curr Pollut Rep 5:353–377. https://doi.org/10.1007/s40726-019-00125-4

Kalmykova Y, Sadagopan M, Rosado L (2018) Circular economy—from review of theories and practices to development of implementation tools. Resour Conserv Recycl 135:190–201

Karunarathne AY, Lee G (2022) How do urban reciprocal support network legacies matter to improve the resiliency of urban informal livelihoods? Sustain Cities Soc 77, Article 103528. https://doi.org/10.1016/j.scs.2021.103528

Kirchherr J, Reike D, Hekkert M (2017) Conceptualizing the circular economy: an analysis of 114 definitions. Res Conserv Recyc 127:221–232

Kirchherr J (2021) Towards circular justice: A proposition. Resour Conserv Recycl 173:105712. https://doi.org/10.1016/j.resconrec.2021.105712

Kirui D (2020) Ex-criminal youth help Africa's largest slums go green and clean [WWW Document]. Euronews

Korsunova A, Halme M, Kourula A, Levänen J, Lima-Toivanen M (2022) Necessity-driven circular economy in low-income contexts: How informal sector practices retain value for circularity. Glob Environ Chang 76:102573

Lee J, Taherzadeh O, Kanemoto K (2021) The scale and drivers of carbon footprints in households, cities and regions across India. Glob Environ Change 66:102205. https://doi.org/10.1016/j.gloenvcha.2020.102205

Lyons C (2016) Urban informality: the production of informal landscapes of musical performance in Sydney. Doctoral thesis: University of Sydney, 1–124.

Marini M (2021) African Cities: Is there Space for Circularity? Main facts, trends and case studies on African urban Circular Economy. Fondazione Eni Enrico Mattei. Italy

Millar N, McLaughlin E, Börger T (2019) The circular economy: swings and roundabouts? Ecol Econ **158**:11–19

Mies A, Gold S (2021) Mapping the social dimension of the circular economy. J Clean Prod 321:128960

Moosmann L (2021) Requested by the ENVI committee. The COP26 climate change conference. https://www.europarl.europa.eu/RegData/etudes/STUD/2021/695459/IPOL_STU(2021)695459_EN.pdf

Mukama T, Ndejjo R, Musoke D, Musinguzi G, Halage AA, Carpenter DO, Ssempebwa JC (2016) Practices, concerns, and willingness to participate in solid waste management in two urban slums in central Uganda. J Environ Public Health, p 7

Murray A, Skene K, Haynes K (2017) The circular economy: An interdisciplinary exploration of the concept and application in a global context. J Bus Ethics 140(3):369–380. https://doi.org/10.1007/s10551-015-2693-2

Murungin N (2021) Vertical farming: Ugandan company develops solution for urban agriculture. Agriculture. https://www.agritecture.com/blog/2021/2/11/vertical-farming-ugandan-company-develops-solution-for-urban-agriculture

Njeri M (2020) From dumpsites to gardens: youth give Dandora facelift. Image Changer. https://www.the-star.co.ke/news/big-read/2020-03-17-from-dumpsites-to-gardens-youth-give-dandora-facelift/a

Okyere SA, Kita M (2016) See, this is a very good place; we are doing many things: resident activities and satisfaction in abese informal settlement, La. J. Sustain. Dev. Afr.18(2):77–100

Okyere SA, Frimpong LK, Diko SK, Abunyewah M, Kita M (2021) Situating everyday urban struggles within the context of the SDGs in an informal settlement in Accra, Ghana. In Sustainable Urban Futures in Africa (Eds) Addaney, M. Cobbinah, P. B. (pp. 265–287). Routledge.

Okyere SA, Frimpong LK, Diko SK, Mensah SL, Pedrosa ELJ (2023) Informal settlements and pro-poor urban planning in African cities: the Luanda Socio-Historical Narrative. In: Cobbinah P, Gaisie E (eds) Reimagining Urban Planning in Africa. Cambridge University Press (In press)

Orayeva JB (2021) COP26: IAEA, UNESCO call for stronger recognition of ocean acidification. IAEA. London. https://www.iaea.org/newscenter/news/cop26-iaea-unesco-call-for-stronger-recognition-of-ocean-acidification

Padgham L, Horne R, Singh D, Moore T (2014) Planning for sandbagging as a response to flooding: a tool and case study. Aust J Emerg Manag 29:26–31

Pati D, Homma R, Iki K (2016) Proposing the use of plastic bottle waste for low-cost housing under government guidelines in India. Am J Eng Sci Technol Res

Quaye I, Amponsah O, Azunre GA, Takyi SA, Braimah I (2022) A review of experimental informal urbanism initiatives and their implications for sub-Saharan Africa's sustainable cities' agenda. Sustain Cities Soc 103938

Rigg J, Oven K (2015) Building liberal resilience? A critical review from developing rural Asia. Glob Environ Change 32:175–186. https://doi.org/10.1016/j.gloenvcha.2015.03.007

Satterthwaite D, Archer D, Colenbrander S, Dodman D, Hardoy J, Mitlin D, Patel S (2020) Building resilience to climate change in informal settlements. One Earth 2(2):143–156

Sawa P (2014) Cheap, green solar bottles light up Kenyan slum. Thomson Reuters Foundation. https://news.trust.org/item/20140221113428-oxrbh/. Accessed 16 Feb 2021

Sesay I (2011) In Africa's largest slum, a cooker that turns trash into fuel [WWW Document]. CNN

Sietchiping R, Yoon H-J, Director I (2010) What drives slum persistence and growth? Empirical evidence from sub-Saharan Africa. Int J Adv Stud Res Africa 1:1–22

Somé J (2018) Overview of Industrial Policy and Institutions in Africa. Researchgate. net/profile/JusteSome/publication/353665550_Overview_of_Industrial_Policy_and_Institutions_in_Africa/links/61094893169a1a0103d72331/Overview-of-Industrial-Policy-and-Institutions-in-Africa.pdf. Accessed 28 June 2022

Tahulela AC, Ballard HH (2020) Developing the circular economy in South Africa: challenges and opportunities. In: Sustainable waste management: policies and case studies: 7th IconSWM—ISWMAW 2017, Volume 1, pp 125–133

Tehlova A (2019) Changing faces competition. Mobilizing citizens to reclaim public spaces in Nairobi. J Public Sp 4:61–86. https://doi.org/10.32891/jps.v4i3.1221

UN-Habitat (2022) World cities report 2022: Envisaging the future of cities. Nairobi, Kenya

UN-Habitat (2016) Slum almanac 2015/2016: tackling improvement in the lives of slum dwellers. UN-Habitat, pp 1–96

United Nations (2015) Chapter 1 crowded with people, not dense with capital. https://documents1.worldbank.org/curated/en/854221490781543956/122290272_201711346052345/additional/113851-PUB-PUBLIC-PUBDATE-2-9-2017.pdf

Walker K (2015) Sandbag housing solution opens doors for home ownership in Nakuru [WWW Document]. Design Indaba. https://www.designindaba.com/articles/creative-work/sandbag-housing-solution-opens-doors-home-ownership-nakuru. Accessed 15 Jan 2021.

Wamboga-Mugirya P (2019) Land-independent production practices raise incomes in Uganda [WWW Document]

Weimann A, Oni T (2019) A systematised review of the health impact of urban informal settlements and implications for upgrading interventions in South Africa, a rapidly urbanising middle-income country. Int J Environ Res Public Health 16(19):3608

World Health Organisation (2016) Pollution: ambient air and, a global assessment of exposure burden of disease. https://apps.who.int/iris/handle/10665/250141 *analysis of 114 definitions.* Utrecht: Innovation Studies Group, Copernicus Institute of Sustainable Development, Utrecht University

United Nations (2015) Millennium development goals and beyond 2015. Goal 7, Ensure environmental sustainability (p. 1). UN

United Nations Economic and Social Commission for Asia and the Pacific (2017). Sustainable development benefits of integrated waste management: Integrated resource recovery centers 51

Stephen Leonard Mensah is a Ph.D. candidate in Urban Affairs at the University of Memphis. He is a community-focused interdisciplinary scholar with research interests in urban and community sustainable development from multidimensional perspectives, including climate urbanism, housing, circular economy, residents' everyday experiences, and policy responses to inequities in local environments. He is particularly interested in exploring these themes using the lens of inclusive, resilient, and sustainable development and the application of varying research methods.

Louis Kusi Frimpong holds a Ph.D. in Geography and Resource Development from the University of Ghana. His research interests are in the area of urban sustainability and multiple aspects of urbanity. Louis has researched and published in areas such as urban safety, urban livelihoods, lived experiences of informal settlement dwellers, urban environmental problems, place making, walkable cities, community-based development, gender and development, and higher education. Dr Louis Kusi Frimpong is currently a Lecturer at the Department of Geography, University of Environment and Sustainable Development, Somanya.

Shine Francis Gbedemah is a Senior Lecturer in the Department of Geography and Earth Sciences, University of Environment and Sustainable Development, Ghana. His area of specialization is in Climate change, Development, and Sustainability Science. He holds a Ph.D. in Human Geography from the University of Leicester, UK. Dr. Gbedemah has a passion for promoting indigenous environmental management practices and their dissemination to for wider policy impact. He teaches various courses like introduction to rural development, introduction to environmental management and corporate environmental management.

Circularizing Livelihoods: Transforming Agricultural Residues to Electricity in Low-Income Peri-Urban Areas of Uganda

Nataliya Apanovich

Abstract This paper addresses the missing link between the circular economy and energy production from agricultural residues in central Uganda. Using wasted agricultural residues for decentralized electricity generation has the potential to sustainably address some of the biggest development challenges in the Global South. To understand which residue types are wasted, the author surveyed smallholder farmers in Masaka District, Uganda. Using the percent of each residue type wasted and the existing gasification technology, it was determined that the region's energy potential represents 14% of the existing electricity capacity of Uganda. The results are interpreted within the agricultural and socioeconomic context of the studied area. Finally, recommendations are made for the realization of a circular economy in Uganda and other African countries with similar patterns of resource generation and use.

Keywords Agricultural residue · Electricity · Energy poverty · Circular energy production

1 Introduction

With limited access to modern forms of energy, many people in Sub-Saharan Africa (SSA) still rely on scarce firewood for cooking, heating, and lighting, often burning it in enclosed spaces. This leads to deforestation, air pollution, and health problems (Jung and Huxham 2018; Kebede et al. 2015; Semenya 2020). Access to modern forms of energy, such as electricity, remains one of the most important determinants of socioeconomic development in Africa (Acheampong et al. 2021). Access to electricity means access to education, income, and food (Blimpo et al. 2020; Candelise et al. 2021; Riva et al. 2018).

It is estimated that only 46% of the population of SSA has access to electricity (IEA et al. 2021). This is representative of Uganda, where only 41% of the population is

N. Apanovich (✉)
School of Landscape Architecture and Planning, University of Arizona, Tucson, AZ, USA
e-mail: apanovich@arizona.edu

S. A. Okyere et al. (eds.), *Urban Slums and Circular Economy Synergies in the Global South*, Advances in 21st Century Human Settlements,
https://doi.org/10.1007/978-981-99-9025-2_8

connected to the main grid (World Bank 2020). Rural electrification barely exists (Lo and Kibalya 2023). Uganda produces enough electricity to electrify more than 41% of its population. Its installed capacity exceeds peak load by 67% (Wabukala et al. 2022). Still, due to nonexistent transmission and distribution infrastructure, excess electricity cannot be consumed domestically and is sold to neighboring countries.

For a household to connect to the main grid, a few things need to happen. First, the household needs to be located close to the distribution powerlines. If so, the household needs to pay inspection fees for the site to be assessed for suitability and then, if approved, connection fees. These fees amount to approximately $700 (UMEME 2023) and are comparable to Uganda's GDP per capita, which in 2021 was $884 (World Bank 2021).

Residential electricity costs around 0.14 ¢/kWh (UMEME 2023), almost as much as in the U.S. (EIA 2022). In addition, connectivity to the grid doesn't guarantee interrupted service. Uganda experiences frequent power outages due to weather events, vandalism, and mismanagement (Wabukala et al. 2022). In times of interruption and to cut expensive electricity costs, Ugandans supplement electricity with other forms of energy, such as charcoal and firewood. Finally, when looking at the spatial distribution of electrification, it becomes evident that the less developed and thus more isolated Northern and the lake regions will most likely not get connected to the main grid soon (Shirley et al. 2021). This leaves millions of Ugandans without access to electricity. This points to a need for a different approach to rural electrification in Uganda.

The evidence suggests that locally developed solutions tend to be cheaper, more accepted, long-lasting, and sustainable (Borelli et al. 2020; Flint and Meyer 2019). Rural Ugandans need a localized model of electrification that can reflect their energy needs, available resources, and challenges. Although agriculture remains the backbone of rural economies with many residue types often perceived as "waste", the potential transformation of agricultural waste to energy remains underexplored. Additionally, the research and praxis around waste-energy discourse have not been positioned within the Circular Economy (CE), despite the immense potential it offers for the social, economic, and environmental improvement of both urban and regional areas.

By relying on the CE framework, electricity can be generated using locally available resources that otherwise would be wasted. Rural communities play a crucial role in efficiently using natural capital as they understand it more closely than any other community (Blades et al. 2017). In CE, resources are kept in use for as long as possible, with the maximum value extracted from them, and materials and resources are recovered and regenerated at the end of each service life. CE relies on three main principles: (1) to preserve and enhance natural capital, (2) to optimize resource yields, and (3) to foster system effectiveness (Velenturf and Purnell 2021). By generating electricity from agricultural waste, rural communities can extract value from a renewable resource that otherwise becomes a source of pollution. Additionally, by employing efficient conversion methods, electricity output will be optimized, leading to the preservation and enhancement of natural capital such as woody biomass and soil.

Literature review (Mutezo and Mulopo 2021; Romero-Castro et al. 2023) shows that CE can enable renewable energy uptake and transitioning, which is necessary for sustainable development. Renewable energy resources and technologies have three main advantages over other energy sources. First, they have a much less environmental impact. Second, if used in appropriate and efficient applications, renewable energy resources can provide a reliable and sustainable energy supply almost indefinitely. Third, they favor power system decentralization and locally applicable solutions (Dincer 2000). These renewable energy benefits mirror the three principles of CE. Thus, it is logical to conclude that energy production from such a renewable resource as wasted agricultural residues has the potential to advance the principles of CE. This model could become a sustainable pathway to rural electrification in an energy-poor country like Uganda.

While many studies link CE and renewable energy, few studies exist on the connection between CE and renewable energy in Africa (Table 1) (Mutezo and Mulopo 2021). Some are centered around using solar energy for water treatment (Falk et al. 2022), while others focus on landfill gas for electricity generation (Mbazima et al. 2022). After identifying only 93 studies and reviewing them on the link between CE and the generation of electricity from agricultural residues in Africa, the author didn't find any substantive discussion on the link. Most of the studies were either global in nature or outside of Africa. The ones that did situate themselves on the continent were focused on the use of agricultural residues for the production of bioplastics (Hierro-Iglesias et al. 2022), biostimulants (Fertahi et al. 2023; Kasinath et al. 2021; Meghana and Shastri 2020), biogas (Surroop et al. 2019), and a cooling system (Alammar et al. 2022). One study discussed electricity potential from biomass in Morocco, emphasizing the techno-economic analysis (Mana et al. 2021). The rest focused on conversion technologies (Khoshnevisan et al. 2021; Plaza et al. 2019; Shahbaz et al. 2021; Sharma et al. 2020). These further cement the significance of this work and the urgency of highlighting the potential role of wasted agricultural residue in energy generation within the emerging CE discourse in Africa generally.

Small-scale, decentralized power generation from wasted agricultural residues can be one of the ways to rural electrification in Africa (Arranz-Piera et al. 2018). Such waste-to-energy technologies as anaerobic digestion, gasification, and combined heat

Table 1 Review and research article results using Boolean operators and wildcards

Search strings	Scopus	Science direct
Period: 2010 to 2023	Search results	
Circular economy + renewable energy	861	5,529
Circular economy + Africa	168	2,286
Circular economy + renewable energy + Africa	6	891
Circular economy + renewable energy + Africa + electricity	2	604
Circular economy + agricultural residues + electricity + Africa	0	93

Search date: February 20, 2023

and power (CHP) have already been employed in Uganda (Table 2). While anaerobic digestion is used more for cooking than electricity production by rural households, gasification, and CHP can provide electricity on variable scales. This study focuses on the gasification technology due to its operational simplicity and proven success in Uganda (Okello et al. 2013).

Agricultural residue production in Uganda is substantial but the percentage that can be realistically and sustainably used for energy purposes is not well established. It is important to differentiate between theoretical and technical residue availability. Theoretical residue potential assumes 100% residue recoverability while technical potential assumes that only a fraction of residues can realistically be recovered. Additionally, it is important to consider competing uses of residues. Sustainability concerns of using residues for energy production should also be addressed.

To address the missing link between CE and energy production from agricultural residues, the author conducted farmer surveys in central Uganda to identify the percent of each residue type wasted and estimate electricity potential using the gasification technology with proven success. The results are interpreted within the

Table 2 Agricultural waste-to-energy technology applications and barriers to adoption in Uganda

Technology	Slow pyrolysis	Gasification	Anaerobic digestion	Combined heat and power (CHP)
Application	Charcoal	Electricity generation	Cooking gas	Electricity generation
Description	The traditional earth-mound kiln is the dominant type of carbonization technology in Uganda and is mostly used for charcoal making (Okello et al. 2013)	Muzizi tea estate and Tiribogo village both use gasification technology for electricity generation. One uses bagasse while the other relies on maize cobs and coffee husks (Mangoyana and Smith 2011)	The most commonly used type of bio-digester is the fixed-dome design. It is estimated that about 500 biogas digesters have been installed with less than 50% still being operational (Walekhwa et al. 2009)	Bagasse, a source for CHP, is produced in large quantities, mainly in three sugar factories: Kakira Sugar Works, Kinyara Sugar Works, and Sugar Corporation of Uganda Limited. The power generated from the factories is used not only internally for sugar processing but is also sold to the national grid (Adeyemi and Asere 2014)
Barriers	Low efficiency, losses during production and transportation stages (Mwirigi et al. 2014)	High cost of technology	Limited technical skills for installation, operation, and maintenance; high capital costs	Used on an industrial scale only (Kaijuka 2007)

agricultural and socioeconomic context of the studied area. Finally, recommendations are made for the realization of CE in Uganda and other African countries with similar patterns of resource generation and use.

2 Methodology

This study represents a mixed method research. Qualitative data (types and uses of agricultural resides) were generated by the author through farmer surveys, while the quantitative data (the annual amount of each crop produced in the region and technical performance of Tiribogo gasifier) came from the Uganda Bureau of Statistics website and literature.

2.1 Study Area: Central Uganda

This study focuses on central Uganda for multiple reasons. With 10 million people, this region is one of the most populous and consists of 22 districts (Mubende, Mityana, Luwero, Kyankwanzi, Mukono, Kayunga, Nakasongola, Nakaseke, Masaka, Kalangala, Buikwe, Kalungu, Lwengo, Mpigi, Kampala, Bukomansimbi, Gomba, Butambala, Buvuma, Wakiso, Kiboga and Kassanda). Annual rainfall ranges between 1,100 and 1,200 mm and comes in two rainy seasons, which allows for two growing seasons per year. Temperatures range between 10 and 30 °C. The topography of the region is characterized by hills and ridges with an average altitude of 1,150 m above sea level (Masaka District 2023). The dominant coffee-banana crop production system is supplemented with cassava (*Manihot esculenta* Crantz), sweet potatoes (*Ipomoea batatas* [L.] Lam), beans (*Phaseolus vulgaris* L.), groundnuts (*Arachis hypogaea* L.), and maize (*Zea mays* L.).

High population density and growth have led to agricultural land fragmentation, cropping intensification, deforestation, rainfall variability, and decreased soil fertility. All of these factors make the central region a good candidate for decentralized, small-scale electricity generation from agricultural residues.

2.2 Sampling

Farmer surveys were conducted during September–October 2015 in Masaka District, covering six sub-counties (Bukakata, Mukungwe, Buwunga, Kabonera, Kyanamukaka, Kyesiga) and one division (Katwe-Butego). In total, 150 smallholder farmers were surveyed, representing 92 villages and 30 parishes.

District-level administrative units consist of sub-counties, parishes, and villages. To choose 150 farmers, five villages were randomly chosen per parish. Then one to six farmers were identified in each village, with an average of two households per village. Farmers were identified from either farmer training records kept by local extension services or by village leaders. A list of all potential farmers (approximately 500 farmers) was created and then used to randomly select 150. In order to capture cross-village variation, the number of farmers surveyed per village was limited to a maximum of six.

Farmers surveyed represented a range of locations and farming systems. The majority (66%) of those surveyed were women aged 30 or older. Most farming households had four to seven members. Most farm sizes were between one to two hectares, with an average size of 1.61 ha. Seven or eight years of primary schooling was the most common educational level achieved. Only 32 percent of farmers surveyed had access to electricity.

2.3 *Tiribogo Power Plant*

To estimate how much electricity can be produced from agricultural residues, the author is using the technical performance of a gasifier and electrical generator located in Tiribogo village, one of the many agrarian communities in central Uganda. This gasifier was chosen due to its successful continuous supply of electricity to a rural community. The Tiribogo plant was installed by Swedish Pamoja Clean Tech AB in 2012 (Pamoja Cleantech 2017) with financial help from Nordic Climate Facility (NCF 2016). The gasifier, Husks Power System, was manufactured in India (Okure et al. 2014).

The plant uses maize cobs supplemented with some coffee husks to run a gasifier seven hours a day to make producer gas to operate an internal combustion engine and generator that can produce 32 kW of electricity (Table 3). Cobs are provided for free by the households who are connected to the mini-grid with $15 per ton for transport. Coffee husks are purchased at $40 per ton, including transport (Okure et al. 2014). System electrical efficiency is 21% and 31% for maize cobs and coffee husks, respectively (Okure et al. 2014). Fuel consumption is 0.88 and 0.74 kg/kWh for maize cobs and coffee husks, respectively. Land was provided free of charge by the chief of the community.

More than half of 200 households could afford to connect to the mini-grid and pay for electricity. Biochar, a by-product, is used to make briquettes for cooking by a local women's group. Some of the challenges with the plant include irregular power generation due to variations in biomass supply, the need for an automatic system to feed in maize cobs, and class tensions in the village due to unequal connection opportunities among households (Eder et al. 2015).

Table 3 Techno-economic summary of Tiribogo village gasifier in central Uganda

	Tiribogo power plant
Company name and origin	Pamoja Clean Tech, Sweden
Project location	Tiribogo village, Central Uganda
Project financing	Nordic Climate Facility through Nordic Development Fund
Capacity	32 kW
Biomass conversion method	Gasification
Type of biomass	Maize supplemented with coffee husks
Annual requirements of biomass	7 tons of biomass
Cost of biomass	Maize cobs are provided for free by the community plus $15 per ton for transport; coffee husks are purchased at $40 per ton including transport
Bio-product	Electricity
By-product	Biochar
Uses of bio-product	- Illuminating agricultural storage; local trade; study and work
Unit price of bio-product	$0.259/kWh (households)
Challenges	Wastewater disposal; no automatic system to feed in maize cobs; access to electricity creates class tensions in the village
Total investment	$81,480

2.4 Electricity Calculations

The author identified major crops grown in central Uganda by surveying farmers. Their annual production was identified by reviewing agricultural production reports on Uganda Bureau of Statistics website (2019). Banana, beans, cassava, maize, and sweet potato are the most common crops and are often grown on a small scale. Coffee and sugarcane are cash crops and can be grown on a commercial scale.

Residue uses of these crops were recorded from farmer surveys. Not every farmer grew all of the above-selected crops; therefore, the number of observations for each residue type differed. Surveys were performed in one district (Masaka District). However, due to the commonality in agricultural and cultural systems, the author assumes that the rest of the farmers in central Uganda used the identified residues for similar purposes.

This study could not capture uses of agricultural residues coming from processing facilities such as sugarcane factories and coffee processing plants as these facilities were not part of the study. Therefore, to assume that all residues from cash crops are retained by the farmer would be inaccurate. The analysis includes these residues in the energy estimate but also discusses how the estimate would be different if they were not included. In order for a residue to be a valid feedstock for local electricity production, it has to be retained by the farmer. Therefore, all residues that are sold with the crop should be discounted from electricity potential.

The author applied the residue-to-crop ratio to get the theoretical amount of residues available from each crop (Table 4):

$$R = C_p \times \text{RPR} \quad (1)$$

where R is the agricultural residue potential (metric tons per year), C_p is the crop production (metric tons per year), and RPR is the residue-to-crop ratio (tons of residue per ton of crop).

The author used two sets of recoverable fractions to estimate electricity potential. The first set of recoverable fractions was identified through a literature review. The author used the recoverable fractions as recommended by Cornelissen et al. (2012)—the realistically recoverable amount of power. In this estimate, the average

Table 4 Agricultural residue uses by farming households in central Uganda (Masaka District)

Residue	Residue uses (%)							
	Number of farmers who grew the crop (n)	Livestock feed	Fuel for cooking	Mulch	Burned without energy extraction	Wasted	Given to neighbor with livestock	Livestock bedding
Banana peelings	138	76	0	16	0	1	7	0
Bean residue	131	5	0	90	1	3	1	0
Cassava peelings	127	69	0	22	0	8	1	0
Coffee husks	19	5	0	74	0	0	0	21
Maize cobs	129	2	92	2	0	2	2	0
Maize stover	129	4	0	92	0	2	2	0
Millet residue	0	NA	NA	NA	NA	NA	NA	NA
Potato peelings	10	30	0	70	0	0	0	0
Sorghum residue	0	NA	NA	NA	NA	NA	NA	NA
Soybean residue	4	0	0	50	25	0	25	0
Bagasse	10	10	0	70	20	0	0	0
Sweet potato peelings	96	80	0	16	0	0	4	0

percentage of recoverable residues was 40%. These fractions did not account for the fact that these residues are produced in a developing country with limited infrastructure and collection machinery. These estimates were used as a benchmark for electricity production. The second set of recoverable fractions came from farmer surveys conducted by the author.

To reflect competing uses, the author chose the percentage of each residue that was burned or wasted as the alternative recoverable fraction for that residue. For example, 1 and 3% of bean residues were burned and wasted, respectively. Therefore, 4% was used as the recoverable fraction for bean residues.

To estimate electricity potential from two sets of recoverable fractions, the author used the performance of the gasifier that is located in Tiribogo village in central Uganda. Specifically, this gasifier has a fuel consumption of 0.74 kg/kWh for coffee husks and 0.88 kg/kWh for maize cobs (Okure et al. 2014). A study conducted by Okure et al. (2014) estimated these fuel consumption rates to determine which residue type was more suitable for this type of gasifier. The study demonstrated that both residue types were suitable for power production. The gasifier, however, runs mainly on maize cobs due to their availability in the village. A more conservative estimate of 0.88 kg/kWh was used for the rest of the residues since no data exists on fuel consumption for other residue types for this specific gasifier. The plant operates seven hours a day, and it is assumed that it is operational 350 days a year. To estimate electricity potential, the author used the following equation:

$$E = \frac{R \times R_f}{F_c \times HPD \times DPY} \quad (2)$$

where E is recoverable electricity (MW), R is the agricultural residue potential (metric tons per year), R_f is the recoverable fraction (decimal percentage), F_c is fuel consumption (kg of residue per kWh), HPD is the hours per day that electricity is generated from a gasifier-powered generator, and DPY is the days per year that electricity is generated.

3 Results

The study identified seven major uses of agricultural residues: livestock feed, fuel for cooking, mulch, wasting/discarding, burning without energy extraction, giving to neighbors with livestock, and livestock bedding (Table 4). The majority of the residues were used as livestock feed and mulch. Farmers with no livestock gave away peelings from cassava, potato, and sweet potato to the neighbors with livestock. Coffee husks were the only type of residue used for animal bedding. The surveys showed that approximately 9% of all residues were either burned or wasted. These residues came from bananas, beans, cassava, maize, soybeans, and sugarcane, with sugarcane producing the greatest amount of residues (Table 5).

Table 5 Selected crops and annual residue production in central Uganda in 2019

Crop	Crop production (million tons/year) (Uganda Bureau of Statistics 2019)	Residue type	Residue-to-crop ratio (Dasappa 2011; Okello et al. 2013)	Residue production (million tons/year)	Recoverable fractions based on technical potential (Cornelissen et al. 2012)	Recoverable fractions based on % of residues wasted
Banana	1.98	Banana stalk and peels	2.0	3.97	0.25	0.01
Beans	0.10	Bean residue	1.4	0.15	0.35	0.04
Cassava	0.41	Cassava stems and peels	0.4	0.16	0.50	0.08
Coffee	0.10	Coffee husk	1.0	0.10	0.75	0
Maize	0.86	Maize cobs	0.3	0.26	0.35	0.02
Maize	0.86	Maize stover	2.0	1.72	0.35	0.02
Millet	0.02	Millet straw	1.4	0.003	0.35	NA*
Potatoes	0.01	Potato vines and peels	0.4	0.006	0.50	0
Sorghum	0.01	Sorghum stalk	1.4	0.02	0.35	NA*
Soybeans	0.03	Soybean residue	2.7	0.007	0.35	0.25
Sugarcane	0.68	Sugarcane bagasse	2.1	1.43	0.19	0.20
Sweet potato	0.34	Sweet potato vines and peels	0.4	0.13	0.50	0

The surveys indicate that most of the residues have established uses and are not available for energy production. This study did not consider the electricity potential from using 100% of maize cobs even though most of them are burned for cooking, and it is done so using inefficient three-stone stoves. It seems like maize cobs are filling an important niche in the household energy mix, and thus it would be unrealistic to divert them to a different use.

Data for millet and sorghum are not available (NA) because farmers in Masaka District do not grow those crops.

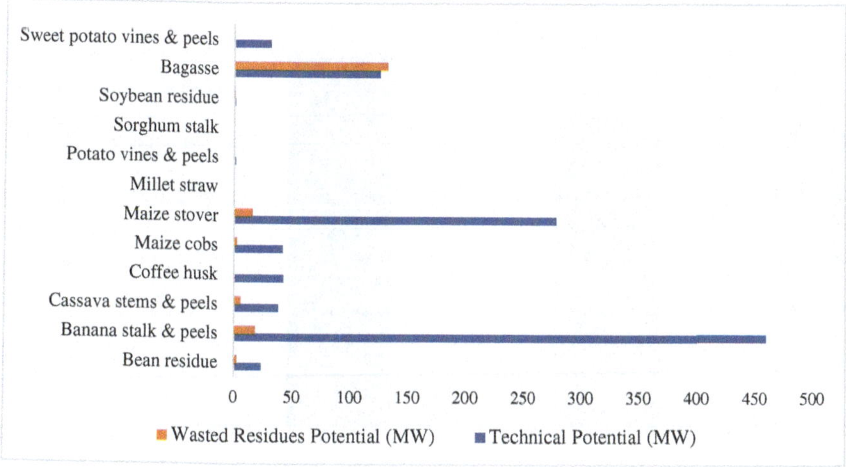

Fig. 1 Electricity potential by residue type

The recoverable fractions established from the surveys and those found in the literature differ substantially. While those found in the literature range from 19 to 75%, the recoverable fractions based on the percent of residue type wasted range from 0 to 20%. Bagasse is the only residue type with the same recoverable fraction as proposed by Cornelissen et al. (2012) and the author of this study. The biggest differences in recoverable fractions come from coffee husks and potato peelings.

Technical electricity estimate produced by applying the recoverable fractions as recommended by Cornelissen et al. (2012) amounted to 1,035 MW, which represents 82% of the current capacity of Uganda (Fig. 2). The greatest energy contributors were banana stalk and peels, maize stover, and sugarcane bagasse. Electricity produced from the wasted residues amounted to 179 MW, representing 14% of the installed capacity. The same residue types contributed most of the energy (Fig. 1).

4 Discussion

The first principle of CE is to preserve and enhance natural capital. Using wasted agricultural residues for energy generation can preserve and enhance forests, water resources, and soil because people will use less firewood for their energy needs. However, it is important to test the soil to ensure that using agricultural residues for energy represents the best possible use of that resource.

The second principle is the optimization of resource yields. Using wasted agricultural residues in such an efficient conversion technology as a gasifier makes it possible to extract maximum value from the residues and use the by-product, biochar, to amend the soil to increase crop yields.

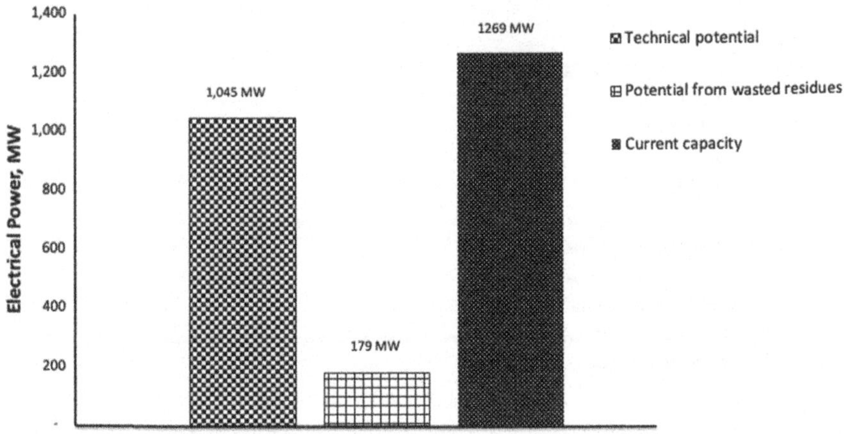

Fig. 2 Comparison of Uganda's current power capacity and electricity produced from technical potential and wasted agricultural residues

The third principle is to foster system effectiveness. While gasifying agricultural residues for energy may seem like an improvement to how residues are currently managed, collecting residues and managing the gasification process might be problematic. Some of these issues arise because agricultural residues are scattered between multiple small plots located far away from each other. Also, as noted by Tiribogo village residents, managing the gasification process requires local expertise. If the technology is foreign and/or managed by an outsider, the residents might feel they need to be more incentivized to learn about and support it. Therefore, seeking technological innovations from within the country or training the community to run the system is essential.

Finally, it is important to note how something so novel as using agricultural residues to produce electricity changes human interactions and perceptions of their community resources and needs. From the case study of Tiribogo village, it is known that unequal access to electricity created class tensions. Those who were connected kept their doors open to show off access to electricity, while those who couldn't connect developed resentment and barriers to communication. It is suggested, therefore, that when introducing such a novel technology to a community, everyone is given an equal opportunity to connect. As some studies on novel agricultural technology adoption show, it is important to consider the characteristics of the farmers (personal, socioeconomic, personality, social networks, status, familiarity with technology), characteristics of the external environment (geographical setting, societal culture, political conditions), and characteristics of agricultural innovations (benefits, costs) (Meijer et al. 2015). This can help gage knowledge, perceptions, and attitudes to ensure long-term technology adoption.

Electricity potential from wasted agricultural residues can increase Uganda's current power capacity by 14% and allow more households in the central region to gain access to electricity. To analyze the importance of this finding within the CE framework, it is necessary to discuss it in the larger agricultural and socioeconomic context and make connections to the CE principles.

Most of the crops in central Uganda are intercropped, with annual and perennial crops often growing together. Due to time and labor constraints and soil quality needs, farmers prefer to leave the residues on the ground as mulch. Usually, this is the easiest way of managing agricultural residues and soil in such a diversified agricultural system. Additionally, on average, each household farms two plots of land with an average plot size of 0.56 ha (Uganda Bureau of Statistics 2019). Almost 40% of all farming households have access to less than 0.5 ha of land. Some plots are accessible only on foot, by bicycle, boat, or motorcycle. Collecting and transporting small amounts of biomass from multiple locations that are not easily accessible can further complicate using agricultural residues for electricity generation.

Another issue is land ownership. Only 78% of all plots in the central region are owned, the rest are rented, or the farmer has only use rights. Of those who claim to own the land, only 52% can produce land tenure documents legalizing the ownership (Uganda Bureau of Statistics 2019). This ambiguity around land ownership can discourage farmers from investing in their land and create conflicts between landlords and tenants (Deininger and Castagnini 2006), all of which can impact people's willingness to use agricultural residues for power generation.

The greatest share of the power potential came from bagasse, banana stalk and peelings, and maize stover. While sugarcane is not a staple crop, bananas and maize are staples, and pests and diseases are increasingly threatening. There is a trend in Uganda where farmers are replacing food crops with cash crops that face fewer biological pressures (Ochola et al. 2022). Coffee is increasingly a crop of choice to replace diseased banana plantations. From farmer surveys, this study found that coffee is used for livestock bedding if farmers get coffee husks back from coffee processing facilities. Other studies found that the soil in coffee fields is more fertile than in banana or intercropped coffee-banana fields because farmers leave more mulch on coffee fields (Bell et al. 2021). With coffee prioritization, there is a threat to energy potential and food security. While women spend more time on agricultural activities, men have more power to manage income from those activities. Such a gendered pattern of spending can lead to financial resource mismanagement and undermine the ability of a household to meet its needs.

Residues from annual crops are more appropriate for electricity production on a local scale based on their availability and suitability for gasification. Annual crop residues require less pre-treatment than perennial crop residues, which tend to be bulkier and thicker. Thus, if farmers continue replacing their annual crops with perennial cash crops, this will compromise their ability to generate enough appropriate biomass types for electricity production.

For small-scale gasification for electricity to be feasible, a community can specialize in one crop, similar to Tiribogo village, where farmers specialize in maize production. Specialization in one crop can ensure a sufficient supply of a feedstock that is uniform in shape and chemical properties, all of which are important for the quality of the final product. However, crop specialization is not possible on a small scale, where farmers grow a variety of crops to meet their dietary needs. Additionally, producing power from a specific residue type may incentivize farmers to grow only one crop which can compromise food security, biological and genetic diversity, and ecosystem services such as soil health. Previous studies have demonstrated that sugarcane-producing communities in Eastern Uganda experienced greater food insecurity than the rest of the country (Lwanga et al. 2015). It was connected to land use change, where important staple crops such as cowpeas, soybeans, finger millet, and yams were replaced with sugarcane (Mwavu et al. 2018). Communities can avoid the negative consequences of practicing a monoculture for energy reasons if they can grow multiple crops that can produce sufficient amounts of residues to meet electricity demand.

The current policies of agricultural modernization in Uganda are switching the focus from small-scale farming to commercial production. If farming becomes more commercial and intensive, greater production of agricultural residues will occur, which, if not properly managed, can lead to waste problems. Linking agricultural and energy sectors can create a more efficient allocation of resources and better environmental and health outcomes.

While it is evident that there are many concerns connected to energy production from small-scale agriculture, it is important to mention that urban or peri-urban areas might be better candidates for decentralized energy production. In more densely populated urban areas, there is a greater production of organic biomass that can be available for energy production. Specifically, restaurants and small food vendors often discard banana and potato peelings without extracting any value from them. Improved transportation systems characterize urban areas and thus can contribute to faster and cheaper ways of organic waste recovery and processing. Also, by recovering organic waste in urban areas, there are no trade-offs between food security and soil quality.

5 Conclusions, Limitations, and Future Directions

This study looked at the role of the potential energy produced from wasted agricultural residues to advance the realization of CE in Uganda. While, wasted residues can meet 14% of the current energy demand using a gasification technology, this method of electrification cannot be described as circular due to the agricultural and socioeconomic realities found in rural central Uganda.

Recovering wasted residues from diversified, small scale, subsistence farming systems characterized by land fragmentation, uncertainty around land ownership and rights, and biological pressures, is not only challenging or even possible but can also put the farmers at risk of food insecurity, soil degradation, and social tensions. It seems that the communities situated in, close, or adjacent to urban areas are better suited for the production of energy from wasted residues in the context of CE. Using agricultural waste from urban or even peri-urban areas is more efficient in terms of residue recovery and eliminates the risks associated with soil quality and food security.

In spite of the contributions of this work, the study has several limitations. First, the author assumed that any community in Uganda can adopt the Tiribogo gasifier. While this gasifier proved successful in one village, it may be unsuccessful in other places. Different communities may have different attitudes and perceptions of modern technologies; therefore, it is essential to conduct more studies to identify communal needs, resources, concerns, and capabilities.

Additionally, the Tiribogo village gasifier represents a unique collaboration between national and international stakeholders, which is unlikely to replicate in other locations. Seeking help and collaboration from within Uganda may be a more viable option for rural communities needing electricity. Second, the author assumed that data collected from one district were representative of the whole region. Third, this study did not consider soil quality concerns, as it would have required an extensive analysis of soils, tillage systems, and climate. Therefore, a more comprehensive approach is needed to estimate which residues and from where can be used for sustainable energy production as well as labor requirements for collection, processing, and pre-treatment of each residue type.

The results of this study suggest a lack of knowledge on the link between wasted agricultural residues and electricity production within the CE framework in SSA. To continue assessing this link, it is suggested that future studies focus on characterizing the quantities, types, and uses of agricultural residues generated in urban and peri-urban areas. Future efforts also need to gage knowledge, perceptions, and attitudes of the people toward new agricultural technologies to ensure long-term adoption. Finally, biomass conversion technologies need to be created from within the country and/or communities to match each area's unique needs and challenges and be affordable enough for everyone to have an equal access to benefit from them.

References

Acheampong AO, Erdiaw-Kwasie MO, Abunyewah M (2021) Does energy accessibility improve human development? Evidence from energy-poor regions. Energy Econ 96:105165. https://doi.org/10.1016/J.ENECO.2021.105165

Adeyemi KO, Asere AA (2014) A review of the energy situation in Uganda. Int J Sci Res Publ 4(1). www.ijsrp.org

Alammar AA, Rezk A, Alaswad A, Fernando J, Olabi AG, Decker S, Ruhumuliza J, Gasana Q (2022) The technical, economic, and environmental feasibility of a bioheat-driven adsorption cooling system for food cold storing: a case study of Rwanda. Energy 258:124560. https://doi.org/10.1016/J.ENERGY.2022.124560

Arranz-Piera P, Kemausuor F, Darkwah L, Edjekumhene I, Cortés J, Velo E (2018) Mini-grid electricity service based on local agricultural residues: feasibility study in rural Ghana. Energy 153:443–454. https://doi.org/10.1016/j.energy.2018.04.058

Bell CA, Namaganda J, Urwin PE, Atkinson HJ (2021) Next-generation sequencing of the soil nematode community enables the sustainability of banana plantations to be monitored. Appl Soil Ecol 166:103999. https://doi.org/10.1016/J.APSOIL.2021.103999

Blades L, Morgan K, Douglas R, Glover S, De Rosa M, Cromie T, Smyth B (2017) Circular biogas-based economy in a rural agricultural setting. Energy Procedia 123:89–96. https://doi.org/10.1016/J.EGYPRO.2017.07.255

Blimpo MP, Postepska A, Xu Y (2020) Why is household electricity uptake low in Sub-Saharan Africa? World Develop 133(105002). https://doi.org/10.1016/j.worlddev.2020.105002

Borelli T, Hunter D, Padulosi S, Amaya N, Meldrum G, Moura De Oliveira Beltrame D, Samarasinghe G, Wasike VW, Güner B, Tan A, Koreissi Dembélé Y, Lochetti G, Sidibé A, Tartanac F (2020) Agronomy local solutions for sustainable food systems: the contribution of orphan crops and wild edible species. Agronomy 10:231. https://doi.org/10.3390/agronomy10020231

Candelise C, Saccone D, Vallino E (2021) An empirical assessment of the effects of electricity access on food security. World Dev 141:105390. https://doi.org/10.1016/J.WORLDDEV.2021.105390

Cornelissen S, Koper M, Deng YY (2012) The role of bioenergy in a fully sustainable global energy system. Biomass Bioenerg 41:21–33. https://doi.org/10.1016/J.BIOMBIOE.2011.12.049

Dasappa S (2011) Potential of biomass energy for electricity generation in sub-Saharan Africa. Energy Sustain Dev 15(3):203–213. https://doi.org/10.1016/J.ESD.2011.07.006

Deininger K, Castagnini R (2006) Incidence and impact of land conflict in Uganda. J Econ Behav Organ 60(3):321–345. https://doi.org/10.1016/J.JEBO.2004.04.008

Dincer I (2000) Renewable energy and sustainable development: a crucial review. Renew Sustain Energy Rev 4(2):157–175. https://doi.org/10.1016/S1364-0321(99)00011-8

Eder JM, Mutsaerts CF, Sriwannawit P (2015) Mini-grids and renewable energy in rural Africa: how diffusion theory explains adoption of electricity in Uganda. Energy Res Soc Sci 5:45–54. https://doi.org/10.1016/J.ERSS.2014.12.014

EIA (2022) Electric power monthly

Falk J, Globisch B, Angelmahr M, Schade W, Schenk-Mathes H (2022) Drinking water supply in rural Africa based on a mini-grid energy system—a socio-economic case study for rural development. Sustainability (Switzerland) 14(15). https://doi.org/10.3390/su14159458

Fertahi S, Elalami D, Tayibi S, Taarji N, Lyamlouli K, Bargaz A, Oukarroum A, Zeroual Y, El Bouhssini M, Barakat A (2023) The current status and challenges of biomass biorefineries in Africa: a critical review and future perspectives for bioeconomy development. Sci Total Environ 870:162001. https://doi.org/10.1016/J.SCITOTENV.2023.162001

Flint A, Meyer zu Natrup C (2019) Aid and development by design: local solutions to local problems. Develop Pract 29(2):208–219. https://doi.org/10.1080/09614524.2018.1543388

Haberl H, Beringer T, Bhattacharya SC, Erb KH, Hoogwijk M (2010) The global technical potential of bio-energy in 2050 considering sustainability constraints. Curr Opin Environ Sustain 2(5–6):394–403

Hierro-Iglesias C, Chimphango A, Thornley P, Fernández-Castané A (2022) Opportunities for the development of cassava waste biorefineries for the production of polyhydroxyalkanoates in Sub-Saharan Africa. Biomass Bioenerg 166:106600. https://doi.org/10.1016/J.BIOMBIOE.2022.106600

IEA, IRENA, UA Statistics Division, The World Bank, and World Health Organization (2021) Tracking SDG7 the energy progress report 2021. www.worldbank.org

Jung J, Huxham M (2018) Firewood usage and indoor air pollution from traditional cooking fires in Gazi Bay, Kenya. Biosci Horiz 11. https://doi.org/10.1093/biohorizons/hzy014

Kaijuka E (2007) GIS and rural electricity planning in Uganda. J Clean Prod 15(2):203–217. https://doi.org/10.1016/J.JCLEPRO.2005.11.057

Kasinath A, Fudala-Ksiazek S, Szopinska M, Bylinski H, Artichowicz W, Remiszewska Skwarek A, Luczkiewicz A (2021) Biomass in biogas production: pretreatment and codigestion. Renew Sustain Energy Rev 150:111509. https://doi.org/10.1016/J.RSER.2021.111509

Kebede A, Chimdi A, Nair AS (2015) Effect of firewood energy consumption of households on deforestation in debis watershed ambo district, Oromia Regional State, Ethiopia. World Appl Sci J 33(7):1154–1163. https://doi.org/10.5829/idosi.wasj.2015.33.07.14586

Khoshnevisan B, Duan N, Tsapekos P, Awasthi MK, Liu Z, Mohammadi A, Angelidaki I, Tsang DCW, Zhang Z, Pan J, Ma L, Aghbashlo M, Tabatabaei M, Liu H (2021) A critical review on livestock manure biorefinery technologies: sustainability, challenges, and future perspectives. Renew Sustain Energy Rev 135:110033. https://doi.org/10.1016/J.RSER.2020.110033

Lo K, Kibalya B (2023) Electric cooperatives and the political economy of rural electrification in Africa: insights from Uganda. Electr J 36:107238. https://doi.org/10.1016/j.tej.2023.107238

Lwanga F, Wanyenze RK, Matovu JK, Garimoi Orach C (2015) Food security and nutritional status of children residing in sugarcane growing communities of east-central Uganda: a cross-sectional study

Mana AA, Allouhi A, Ouazzani K, Jamil A (2021) Feasibility of agriculture biomass power generation in Morocco: techno-economic analysis. J Clean Prod 295:126293. https://doi.org/10.1016/J.JCLEPRO.2021.126293

Mangoyana RB, Smith TF (2011) Decentralised bioenergy systems: a review of opportunities and threats. Energy Policy 39(3):1286–1295. https://doi.org/10.1016/J.ENPOL.2010.11.057

Masaka District (2023) Masaka district

Mbazima SJ, Masekameni D, Mmereki D (2022) Waste-to-energy in a developing country: the state of landfill gas to energy in the Republic of South Africa. Energy Explor Exploit 40(4):1287–1312. https://doi.org/10.1177/01445987221084376

Meghana M, Shastri Y (2020) Sustainable valorization of sugar industry waste: status, opportunities, and challenges. Biores Technol 303:122929. https://doi.org/10.1016/J.BIORTECH.2020.122929

Meijer SS, Catacutan D, Ajayi OC, Sileshi GW, Nieuwenhuis M (2015) The role of knowledge, attitudes and perceptions in the uptake of agricultural and agroforestry innovations among smallholder farmers in sub-Saharan Africa. Int J Agric Sustain 13(1):40–54. https://doi.org/10.1080/14735903.2014.912493

Mutezo G, Mulopo J (2021) A review of Africa's transition from fossil fuels to renewable energy using circular economy principles. Renew Sustain Energy Rev 137. https://doi.org/10.1016/J.RSER.2020.110609

Mwavu EN, Kalema VK, Bateganya F, Byakagaba P, Waiswa D, Enuru T, Mbogga MS (2018) Expansion of commercial sugarcane cultivation among smallholder farmers in Uganda: implications for household food security. Land 7(2):73

Mwirigi J, Balana BB, Mugisha J, Walekhwa P, Melamu R, Nakami S, Makenzi P (2014) Socio-economic hurdles to widespread adoption of small-scale biogas digesters in Sub-Saharan Africa: a review. Biomass Bioenerg 70:17–25. https://doi.org/10.1016/J.BIOMBIOE.2014.02.018

NCF (2016) Nordic climate facility

Ochola D, Boekelo B, van de Ven GWJ, Taulya G, Kubiriba J, van Asten PJA, Giller KE (2022) Mapping spatial distribution and geographic shifts of East African highland banana (Musa spp.) in Uganda. PLoS ONE 17. https://doi.org/10.1371/journal.pone.0263439

Okello C, Pindozzi S, Faugno S, Boccia L (2013) Development of bioenergy technologies in Uganda: a review of progress. Renew Sustain Energy Rev 18:55–63. https://doi.org/10.1016/J.RSER.2012.10.004

Okure MA, Musinguzi W, Wabwire A, Bagenda S (2014) Experience from rural electrification in Uganda: a case study of a Husk powered system in Tiribogo Village

Pamoja Cleantech (2017) New 50kW project in Uganda with agricultural hub-2017

Plaza D, Artigas J, Ábrego J, Gonzalo A, Sánchez JL, Dro AD, Richardson Y (2019) Design and operation of a small-scale carbonization kiln for cashew nutshell valorization in Burkina Faso. Energy Sustain Dev 53:71–80. https://doi.org/10.1016/J.ESD.2019.10.005

Riva F, Ahlborg H, Hartvigsson E, Pachauri S, Colombo E (2018) Electricity access and rural development: review of complex socio-economic dynamics and causal diagrams for more appropriate energy modelling. Energy Sustain Dev 43:203–223. https://doi.org/10.1016/j.esd.2018.02.003

Romero-Castro N, Ángeles López-Cabarcos M, Miramontes-Viña V, Ribeiro-Soriano D (2023) Sustainable energy transition and circular economy: the heterogeneity of potential investors in

rural community renewable energy projects. Environ Dev Sustain. https://doi.org/10.1007/s10668-022-02898-z
Semenya K (2020) Environmental health risks associated with firewood induced volatile organic compounds in Senwabarwana villages, republic of South Africa
Shahbaz M, AlNouss A, Ghiat I, Mckay G, Mackey H, Elkhalifa S, Al-Ansari T (2021) A comprehensive review of biomass based thermochemical conversion technologies integrated with CO_2 capture and utilisation within BECCS networks. Resour Conserv Recycl 173:105734. https://doi.org/10.1016/J.RESCONREC.2021.105734
Sharma S, Basu S, Shetti NP, Kamali M, Walvekar P, Aminabhavi TM (2020) Waste-to-energy nexus: a sustainable development. Environ Pollut 267:115501. https://doi.org/10.1016/J.ENVPOL.2020.115501
Shirley R, Liu Y, Kakande J, Kagarura M (2021) Identifying high-priority impact areas for electricity service to farmlands in Uganda through geospatial mapping. J Agricult Food Res 5:100172. https://doi.org/10.1016/J.JAFR.2021.100172
Surroop D, Bundhoo ZMA, Raghoo P (2019) Waste to energy through biogas to improve energy security and to transform Africa's energy landscape. Curr Opin Green Sustain Chem 18:79–83. https://doi.org/10.1016/J.COGSC.2019.02.010
Uganda Bureau of Statistics (2019) Uganda Bureau of Statistics. Crop production
UMEME (2023) What you need to get connected
Velenturf APM, Purnell P (2021) Principles for a sustainable circular economy. In: Sustainable production and consumption, vol 27. Elsevier B.V., pp 1437–1457. https://doi.org/10.1016/j.spc.2021.02.018
Wabukala BM, Bergland O, Rudaheranwa N, Watundu S, Adaramola MS, Ngoma M, Rwaheru AA (2022) Unbundling barriers to electricity security in Uganda: a review. In: Energy strategy reviews, vol 44. Elsevier Ltd. https://doi.org/10.1016/j.esr.2022.100984
Walekhwa PN, Mugisha J, Drake L (2009) Biogas energy from family-sized digesters in Uganda: critical factors and policy implications. Energy Policy 37(7):2754–2762. https://doi.org/10.1016/J.ENPOL.2009.03.018
World Bank. Access to electricity in Uganda (2020). https://data.worldbank.org/indicator/EG.ELC.ACCS.ZS?locations
World Bank (2023) GDP per capita in Uganda. Retrieved 9 Feb 2023 from https://data.worldbank.org/indicator/NY.GDP.PCAP.CD?locations=UG

Metropolitan Cartography: A Novel Approach for Assessing How New Morpho-Types Solutions Impact the Circular City Agenda. The Ouagadougou Case Study

Antonella Contin and Alessandra Pandolfi

Abstract Global South cities are reclaiming a new metropolitan cultural framework and a structural matrix that binds different urban *morpho-types* to improve its economic and spatial dynamics. Today, new and old ways to use local resources are scaled in time among practices that exist since the *pre-colonial, colonial and post-colonial phases*. To align the theoretical and conceptual grounds of the circular economy with sustainable urban development principles in the Global South, the potential benefits associated with non-planned or low-income settlements in promoting circular and sustainable cities should be investigated by recognising the infrastructure connections between planned (*loti*) and non-planned (*non-loti*) settlements. This also means trying to overcome the classic monocentric city model regarding services, functions and economic attractivity to create more accessible and equitable spatial distributions that consider accessibility for all and facilities diffusion as the main values of a balanced urban structure. It is therefore essential to shape an inclusive planning approach to support citizenship by generating knowledge for a collective action based on a new city model's storytelling through maps. The MSLab's innovative approach introduces an open-source mapping methodology: Metropolitan Cartography, which relates the geographic ground to spatial structures to manage scarce urban resources. We aim to showcase the effectiveness of an inclusive planning approach that utilises a new city model's storytelling through open-source maps as a tool. By doing so, we seek to empower metropolitan citizens to generate knowledge for collective action.

Keywords Metropolis · Cultural framework · Structural matrix · Green-grey infrastructure · Metropolitan cartography · Slum-CE synergies

A. Contin (✉) · A. Pandolfi (✉)
Politecnico di Milano, Dipartimento di Architettura e Studi Urbani, Milan, Italy
e-mail: antonella.contin@polimi.it

A. Pandolfi
e-mail: alessandramaria.pandolfi@polimi.it

S. A. Okyere et al. (eds.), *Urban Slums and Circular Economy Synergies in the Global South*, Advances in 21st Century Human Settlements,
https://doi.org/10.1007/978-981-99-9025-2_9

1 A New Perspective on the Global South Urban Development

This chapter is meant to highlight the need for a theoretical and conceptual foundation on how circular economy practices converge with the principles of sustainable urban development in Global South contexts. We do this by exploring the positive role of non-planned/low-income settlements in the agenda for circular and sustainable cities (Korhonena et al., 2018). This requires moving away from the classic model of a monocentric city towards a more accessible and equitable spatial distribution that considers accessibility for all and facilitates diffusion as the main values of a balanced urban structure.

Our research (Contin et al. 2013) aligns with the formal and structural aspects of urbanity at the metropolitan scale. To improve urban development and its cultural understanding, we propose a better definition of the structural model in terms of physical dimension and identify gaps in urban development. Nevertheless, our context is not only about urban spaces but also concerns all the interactions between the physical dimension and *drivers of change* (Strategic Plan 2020–2023—UNHabitat 2019), which must determine sustainable practices, but also their supply, demand and price (Mayo and Angel 1993). A robust economic attraction related to the infrastructure system can help legalise non-planned settlements by providing better services. Though the proposal seems overly ambitious, we are strongly convinced that, going beyond hygiene and welfare goals, our approach could trigger a productive revolution in the metropolis by integrating green and grey infrastructure towards regenerative agriculture into the *new urban gradient of formalities.*

In view of this, our contribution aims to define how to support citizenship and generate knowledge for collective action, through an inclusive planning approach, based on a new city model's storytelling through maps. Our research is focused around the rapidly developing metropolises located in countries within the southern latitudes. We explore how the surrounding context influences metropolitan governance and examine the non-linear and evolving relationships between times and values. Rather than relying on absolute planning scenarios, we adopt a proportional approach that demands thoughtful consideration (Contin et al. 2022a). The MSLab's innovative approach introduces an open-source mapping methodology: Metropolitan Cartography (MC), which relates the geographic ground to the territorial organisation to manage scarce urban resources (Contin et al. 2021b). In the Metropolitan Cartography, we aim for a functional integration of the past and present, creating a sense of resonance and common belonging. The maps in Metropolitan Cartography serve as tools to support design choices in metropolitan architecture projects, to foster relevant relationships and shared values (Fig. 1). By projecting these visions, we can effectively manage the complexity inherent in metropolitan landscapes. In our interpretation, in fact, the concept of heterotopia, the new urban morphotype discussed by Shane (2005), is intimately connected to mapping, which helps us navigate and understand the diverse spatial and social dimensions present within these metropolises.

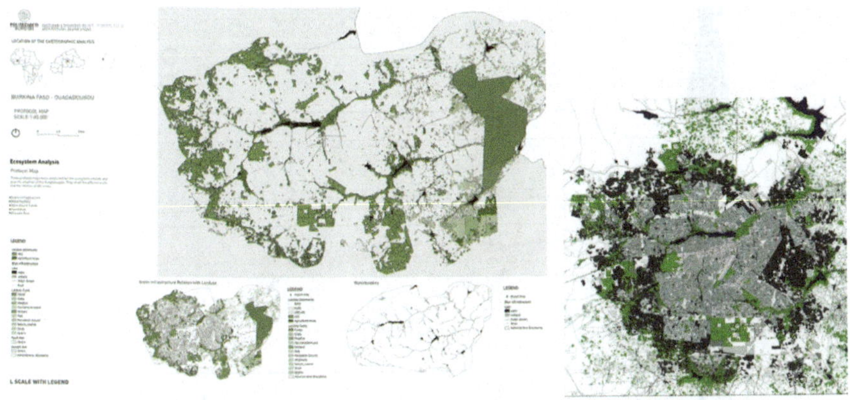

Fig. 1 Burkina Faso. Ouagadougou. Protocol Map. Ecosystem Analysis (MSLab)

The narrative, interpreted in the light of some recent projects implemented in Ouagadougou (Burkina Faso), is based on the possible coexistence between planned urban morphology's western forms and the local non-planned city extension multiplying cellular. That integration model must involve the four metropolitan dimensions: physical, social, economic, and governance, determining which city growth model's infrastructure can transform the non-planned without destroying it, and tracing back to the territory structure resilient points. The notable points-built map allows studying the impacts, defining formal and informal cities and their inhabitants' lifestyle pact (Di Stefano and Lepratti, 2016). It is also the fundamental starting point to promote the non-planned CE synergies, benefiting the whole city (Dantas et al., 2021).

Our research focuses on the non-planned (*non-loti*) settlements on the fringes of the metropolis of Ouagadougou extension due to the constant influx of migrants from rural areas. We aim to develop a new urban morphotype (Rivetta 2007) of planned and non-planned inclusion as an urban–rural linkage pattern. Rather than a rigid formal model, we apply a development model that not only focuses on housing and urban services but considers cities as geographic sites with cultural value, income generation or productivity (Cohen 2009). It consists of both patterns of settlement growth and metropolitan architecture projects (Contin and Ortiz 2015) that suggest possible configurations characterising individual qualities with the *gradient of formality*. According to Pedro Ortiz (Ortiz 2014), the Metro-Matrix is a valuable tool within the practice of Metropolitan Discipline. It comprises a network-like system of axes, including a few primary geographic axes and additional penetrative ones. These axes play a dual role: they determine the development of structural axes and define pivotal points of activity and densification. The Metro-Matrix derives its legitimacy from the logic of historical and geographic settlement patterns, allowing for the identification of interchange nodes, even in peripheral areas of the metropolitan region previously considered distant from the central core (Fig. 2).

Fig. 2 Burkina Faso. Ouagadougou, Metropolitan Architecture Project, Masterplan Özge Tuncay, MSLab, 2023, Master Thesis. Politecnico di Milano

The Metro-Matrix is particularly suitable for non-formal areas often developed through a disjunctive incrementalism approach that fails to capture the complexity of the urban landscape immediately. Thus, the Metro-Matrix offers new possibilities by departing from our usual mental frameworks. Metropolitan Cartography aims to translate intuitive evidence into communicable terms, employing the language of spatial disciplines or other symbolic scripts. Concurrently, the physical execution of the Metro-Matrix descriptions and schemes is integral to the simplification process, shaping our experience just as much as the intuition underlying it. It is not merely about 'codifying' the territory but attributing qualitative value to the points identified through the matrix scheme.

Through meticulous graphological semiology, the maps of Metropolitan Cartography give substance and form to our understanding of the schemes and diagrams derived from the Metro-Matrix. These maps enhance our knowledge of metropolitan dynamics by representing the qualitative 'variations' of each identified position (Fig. 3). They illustrate the precise conditions under which observations can be communicated. Consequently, once recognised, and interrelated, local variations at each position become invariant when examined at the metropolitan scale. The research question addresses the need for a cultural 'jump' that identifies the physical and social infrastructure to implement new relationships among the existing DNA city's elements.

As a new urban morphotype, the metropolitan urban–rural linkage pattern is the primary physical infrastructure for a sustainable social economy as the CE starting

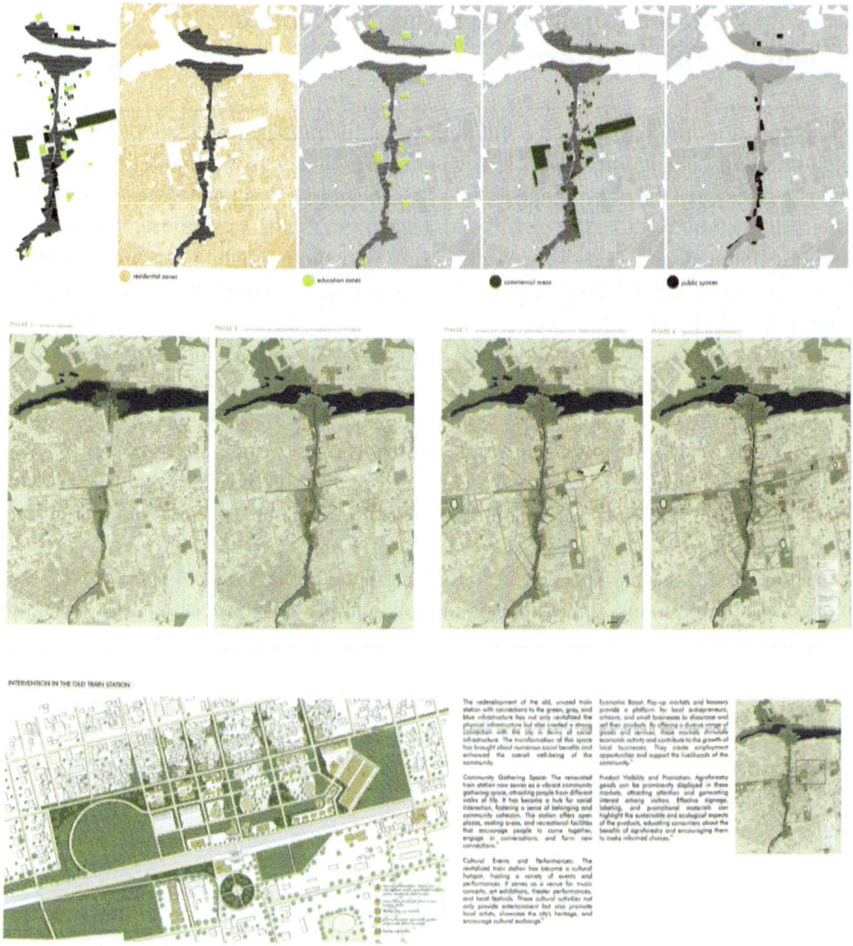

Fig. 3 Burkina Faso. Ouagadougou, Metropolitan Architecture Project, Masterplan (MSLab, Özge Tuncay)

point (Irajpour and Pandolfi, 2023). This indicates the importance of long-lasting relationships between non-planned/low-income settlements, the physical context situation and their inhabitants in determining the sustainable development of Global South cities (Panchal et al., 2021).

2 A New Model for the Integration of Planned and Non-planned Areas

The planned model partly grows and decreases; the non-planned cannot restructure itself internally. So, it is necessary to determine which city growth model's infrastructure can transform the non-planned without destroying it, tracing back to the territory structure resilient points. The notable points-built map allows studying the impacts, defining planned and non-planned city parts and their inhabitants' lifestyle pact. That is the fundamental starting point for promoting the non-planned CE synergies, benefiting the city (Fig. 4).

In his work 'The Ecological Thought', Edgard Morin introduced the idea of the Ecology of Form, which goes beyond just technical or economic efficiency and encompasses the fundamental aspects of life processes and structures. In our study (Contin et al. 2013), we are dedicated to crafting metropolitan architectural forms, including new building form types and land uses that we have designated as 'new urban morpho-types'. Our approach takes a comprehensive view that acknowledges the complex nature of the urban and architectural metropolitan environment, which comprises both planned and unplanned elements. The first step of our project is to understand the elements that make up the Ouagadougou metropolitan city to create

Fig. 4 Ouagadougou. Former airport area project. New Railway Station. New urban morphotype as linkage urban–rural pattern (MSLab & Marcello Longeri, Christodoulos Iosif)

a cultural and physical form that connects to its meaning. The process is morphogenetic, meaning the form is created first and then planning and policy techniques are employed.

The goal is to drive socio-economic development through metropolitan relations and innovation, enhance infrastructure, promote renewable energy and civil economy (Bruni et al. 2019), and improve intermodal transportation while meeting qualitative and quantitative standards. It is also intended to improve and include these settlements with formal ones. In addressing the issue of the appropriate development paradigm for integrating planned and non-planned settlements in metropolitan cities, three key factors must be considered from an architectural and urban design viewpoint: time, the scale of intervention and the models. Besides, it is the physical development that can support a circular economy model (Fig. 5).

Economically speaking, the development of cities and metropolitan areas has often been focused on increasing urban values, including economic and real estate values as well as the quality of life and environmental factors (Pandolfi 2019). However, public finance cutbacks have made it increasingly necessary to involve citizens and private investors in urban development. Strategies, such as creating new functions or promoting the regeneration of public facilities, can attract new potential metropolitan users without supporting unsustainable expansion trends. To achieve a balance between visions, and interests city strategies should start by expressing equality judgments and choosing between multiple policies linked to specific visions, interests and approaches (*ibidem*), including higher attention towards CE, which can become a driver for a new economic model for metropolitan systems.

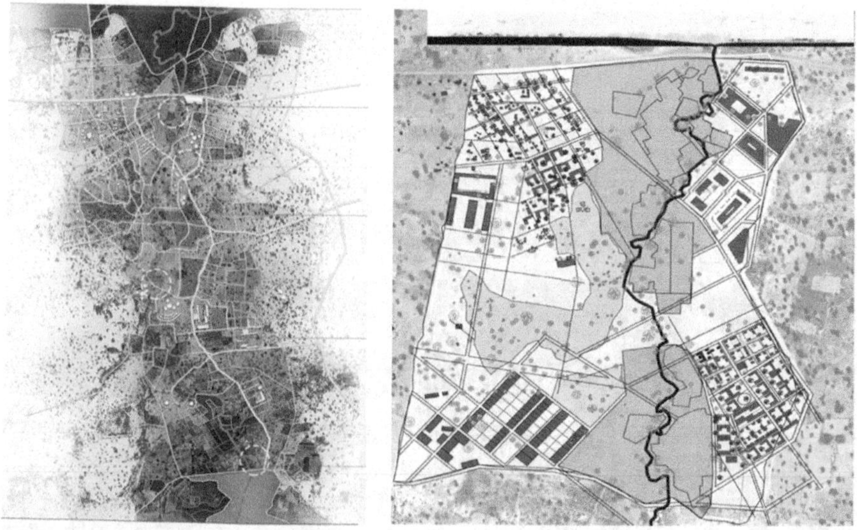

Fig. 5 Ouagadougou. Koubri Green Axis Section (MSLab & Yalda Beigagha, Ilayda Comert, Kubra Ilayda Turan)

The actual metropolitan studies use economic tools to analyse the allocation of economic goods, with non-movable spatialized properties taking a central stage. Urban and metropolitan planners require a large set of skills to evaluate the complexity of urban systems (Fig. 6) and develop sustainable planning strategies (Conconi and Pandolfi 2018). Value assessment is essential in estimating metropolitan assets involving data from technical and analytical disciplines, as well as economic quantification of asset values. Using innovative and reliable techniques in real estate appraisal sciences can have cultural implications beyond economic value.

The need for new tools and applications in urban and metropolitan planning sciences has become more pressing in the last two decades. For many authors, the solution lies in controlling urban rent and adopting innovative approaches to metropolitan planning that consider the interests involved in relevant decisions beyond mere spatial allocation.

Furthermore, in order to understand that non-planned settlements play a significant role in the development of a society, the chapter aims to include the planned and non-planned realms defining a metropolitan urbanity strongly connected to the social and economic ecosystems. Overall, it seems that the issue of urban rent and its relationship with land management regulations, settlement costs, soil classes and the real estate market is still a central concern in urban planning and development (Pandolfi 2019). This concern could be addressed through CE practices. The fact that

Fig. 6 Ouagadougou. Urban Section Landscape showing the connection from Centre to Koubri (MSLab & Yalda SSSBeigagha, Ilayda Comert, Kubra Ilayda Turan)

private developments often benefit from the surplus and advantages related to public city development highlights the need for more equitable policies that redistribute these benefits to the community (*ibidem*). Evaluation and consistent rules are crucial in achieving this goal while ensuring that entrepreneurial initiatives are not overly affected.

For many authors, the solution lies in controlling urban rent and adopting innovative approaches to metropolitan planning that consider the interests involved in relevant decisions beyond mere spatial allocation. The approach is no longer to prosecute informality but to include it in the general development theory for social and economic integration, urban sustainability, and governance equilibrium. The outcome is urban inclusion through a design method that describes and designs urban phenomena related to changes in the scale of city morphologies and architectural types, urban spaces and landscapes. The Green and Grey Infrastructures, Housing areas, Social urban centralities and Productive areas can be designed as a Pattern Language mechanism of metropolitan urbanity development through a Design Guide, providing a framework for the planned and non-planned dialogue. Effective planning and design require coordination between planning, design and metropolitan architecture management.

That approach deals with metropolitan complexity (Contin et al. 2021a). Considering that urban areas make up over 80% of global GDP and over 40% of all GDP in developing countries (WB 2023; McKinsey 2011), we want to link to urban spaces, sites or flows the economic stimuli inset in the city development. The multilateral projects often aim to boost economic activities in cities by financing physical investments, but they lack a spatial and physical focus. These investments are not considered for their potential negative or positive externalities, such as environmental impact or bundled infrastructure services. Essentially, the localisation and design of these projects disregard decades of expert local knowledge on the social, environmental, political and cultural aspects of urban infrastructure and the built environment.

The metropolitan form is created by the placement and integration of large structural elements that define the metropolitan scale and bring urban-style living conditions to suburban areas. A few key infrastructure nodes control the metropolitan territory, shaping its form. For the metropolitan area to be effective, these nodes must be strategically placed to harmonise with green environmental and grey transport infrastructures. Imposing a bottom-up strategy on a metropolitan area is a mistake; rather, it should be developed to fit the territory and sustainably use its assets, such as transportation systems and regional economic or social projects (Ortiz 2014).

The relationship between economics and planning and the balance between efficiency, prosperity and equity can be complex. This conflict results from the impact of economic activities and large infrastructures on the environment and the need for social sensitivity in limiting their impact. This leads to a new *territorialization of the region*. To address this, an integrated model is needed to balance the need for rapid and diverse communication while respecting the environment.

The design of the metropolis must stem from a spatial concept that blends the key strengths of its districts into a cohesive form. After this strategic form is identified, specific projects can be strategically located. Each project must align with the strategic concept as it is applied explicitly to its district, with the concept taking shape through the implementation of the projects.

First, our study focuses on the regional scale and considers geography as the intersection of physical and social cities. We aim to establish a paradigm model, a vision and an idea for orderly city growth to localised new metropolitan functions. All social activities leave physical and symbolic marks. It is essential to consider how these connect to the geographical and historical context of existing sustainable habitats and their physical and spatial patterns established in the past. Additionally, it is essential to consider how these patterns relate to the concept of the metropolis and the idea of global cosmopolitanism (Fig. 7).

The circular economy, as an economic model focused on conserving the socio-economic value of goods by minimising waste and maximising the use of renewable resources (Ellen Mac Arthur Foundation 2012), requires technological advancements and territorial policies implemented effectively. Environmental and Economic factors, including increased product value-added and reduced pollution, can measure it. The design of goods, territorial policies and the scale of implementation all play a role in the success of the circular economy. A development model compatible with the principles of the circular economy is *self-centred development based* on maximising local value production and minimising external dependencies since the concept of the circular economy has roots in the global sensitivity to the negative ecological effects of capitalism. It has developed from studies on the limits of growth and the connection between ecology and economy: "*The circular economy, defined in a stringent sense by the Ellen MacArthur Foundation since 2012, is today a model of interrelation between the phases of the economic process. This model deals mainly with the life cycles of matter, distinguishing it in biological and non-biological (or 'technical'), but from the beginning has called for simultaneous attention to energy, hoping for the replacement of fossil sources with solar sources*" (Lepratti 2021).

3 Circular Economy for the *Loti* and *Non-loti* Settlements Linkage in Ouagadougou

Our pioneering concept revolves around reimagining, planning and reconstructing the way of life in Ouagadougou, considering the country's intellectual capital, particularly concerning the water system (barrages and canals), a well-maintained green-grey infrastructure and agro-forest production (Pagano and Rossi 2009). However, our proposal is rooted in the concept of optimisation, which sets it apart from an approach focused on maximisation. From a metropolitan physical perspective, our proposal explores the possibilities of green transitions to strengthen the informal areas, mitigate negative impacts, fully leverage the potential of environmental, social and economic sustainability, and establish a structure capable of supporting a circular economy strategy.

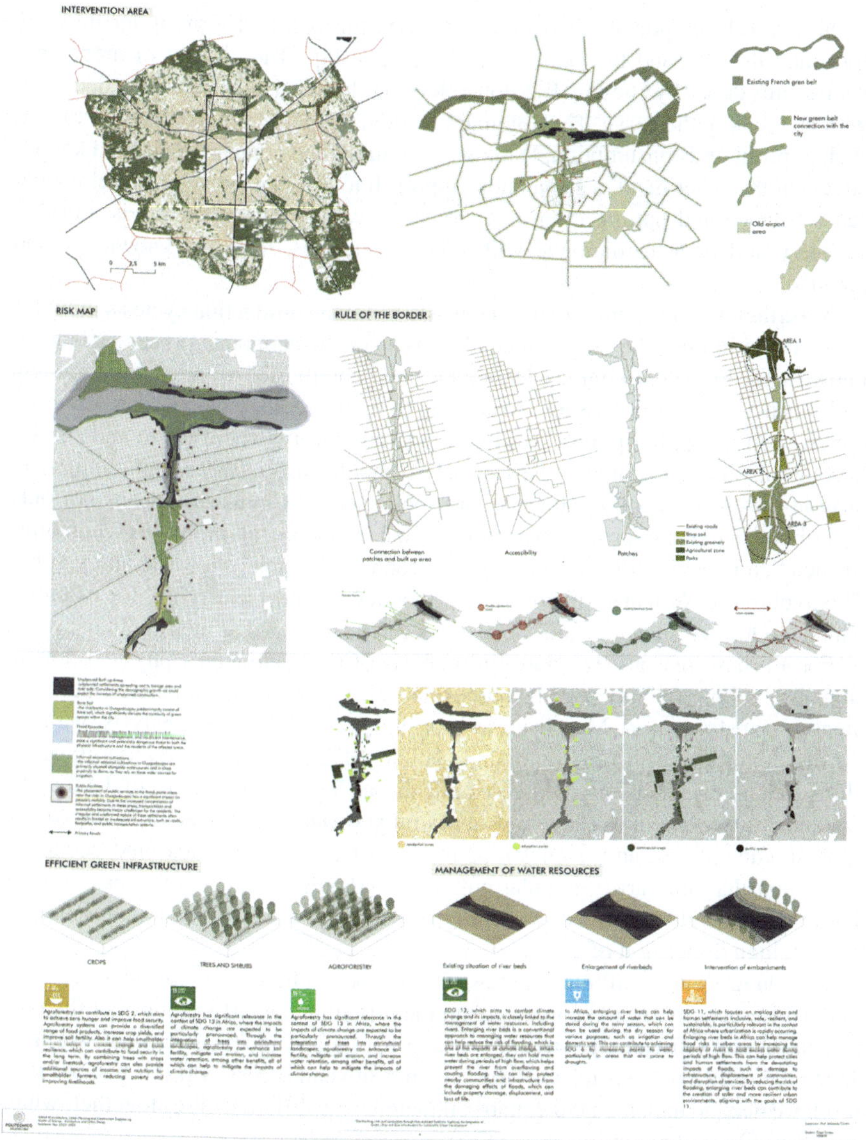

Fig. 7 Ouagadougou. Metropolitan Architecture Project Physical dimension able to sustain a circular economy project (MSLab & Ozge Tuncay)

According to Giordano (Contin et al., 2021a), values play a crucial role in decision-making and evaluation processes. They may be implicit, stem from an individual's worldview and epistemology, and are involved in knowledge construction. Currently, a dominant scientific and utilitarian approach to knowledge tends to separate it from values.

In urban development, values are expressed through transforming territory, its form and the city and the environment's relationship. The concept of monotonic values, introduced by Gregory Bateson, refers to values that only increase or decrease, with no break in the curve. The dominant worldview focuses on maximising values, such as profit and consumption, leading to serialised and compartmentalised knowledge. On the other hand, a sustainable approach to African territorial development seeks balance and optimisation of a variety of values. This approach is transdisciplinary and open to dialogue with other disciplines, technical knowledge and practice.

Nevertheless, the attempt to use the waste from one production cycle as input for another to minimise the environmental impacts of the economic processes is well-known in the local communities. Di Stefano and Lepratti tried to enrich this paradigm with financial and social issues. According to them, ecology emphasises key principles, such as tipping points, resilience, and circularity, where waste and surplus become resources for others. This principle should inspire the economy, particularly in the 'real' economy, finance and social issues. Circularity means not only describing an existing process but also the goal of making all processes (economic, financial and social) efficient, fair and environmentally upcycle waste and surplus. This requires governance of flows at various levels and identifying, for example, which productive, financial and social flows can be managed as metropolitan cycles.

For all these reasons, we start our analysis of Ouagadougou's physical dimensions by looking for the asset resources and their possible maintenance, development or transformation. Moreover, as a background of that first analysis, we took the decolonial perspective (Contin and Galiulo 2022). Given the pressing issues of climate change, increasing socio-economic disparities and the decline of democratic accountability, we aim to show how ancestral knowledge, when acknowledged and applied, can inform, and inspire new models of cultural, political and economic innovation that are part of popular culture and rooted in indigenous cosmovision. These models hold relevance in shaping present-day urban life and addressing the metropolitan dimension of cities.

The Ouagadougou cultural framework must be explicit if we consider the 'non-loti' a different city from the mainstream metropolitan idea. If we consider the non-loti, the actual African city structure where norms and practices coexist, we must define if this difference is related to use, ownership and access or if it is historical. In short, we need to understand the matrix that holds the different cities together, which can shape Ouagadougou today.

Colonial culture precluded locals from the freedom to use the land. The construction and planning practices of the two cities can direct us to search for a code of mediation. Today, new and ancient ways to use the territory resources are scaled in time among practices that date back to the three pre-colonial, colonial and post-colonial phases, the latter of which must be carefully dated to make explicit the intentionality of the nowadays refusal.

Inclusive planning approaches aim to support active citizenship by generating capacity for collective action based on a city's narrative between history, chronicle, social and individual projects. Politecnico di Milano MSLab's innovative approach introduces an instrument as an open-source mapping methodology. The Metropolitan Cartography and the Software Glossary establish how a territory is organised and managed to exploit its scarce resources sustainably. The Glossary outlines the fundamental concepts and concerns that form the basis of maps, which involve recognising significant landmarks that hold enduring and meaningful relationships with one another. It converts memory into value by organising it through keywords, interconnected concepts and symbols, thereby acknowledging the values that originated them.

The map is a non-verbal narrative presenting how the past and the present territory's elements coexist and what hierarchies and dialogue take place between different agents, enhancing the sensitivity of the formal planning process to the dynamics of local needs. Metropolitan Cartography tool as a Built Map made of signs/indexes of relationship among geography, remarkable points, roads and canals indicates places and their images as mnemonic signs of the metropolitan mental map. A map is necessary because it finds the resilience and inclusion of metropolitan citizenship through a deep understanding of the need for metropolitan work.

One of the terms Narrative's meanings is 'Part of the foundations of the judgement in which the course of the trial is set out'. Ours is the narration of territory: how a territory is organised and managed to exploit its scarce resources sustainably. Nevertheless, the narrative of a city lies between history, chronicle and social and individual projects.

'Mapping' is our approach to structuring our judgement on metropolitan city's challenges (Dobbs et al., 2011). We want to narrate how the map of that territory and city is made: it is made up of remarkable points found and made throughout history and the stories of men and women who made the city (Erodoto). The map is, then, a non-verbal narrative that starts with the ability to regulate water, cultivate and raise animals (Cain and Abel, shepherds and gatherers, Eneide and Virgilio), build markets (caravans), create roads, railways and airports today.

The map interprets the reasons for today's project, its plan of reality and not an ideal one. We are interested in combining the concept of decolonial with that of local tradition's resistance. But which tradition? We need to read the idea of the city of the past in an open way that we can define as 'cultural' and, therefore, not only focus on the theme of law and occupation, which leads us to analyse the value matrices only on an economic or de jure basis.

4 The Disciplinary Issue Related to Metropolitan 'Nature'

The cartography tool for the deep knowledge of the 'where' to understand what cannot be erased is our answer. We propose an interpretation of the large barrages and canal system to force the public dimension. A strip model in the manner of Koolhaas absorbs services and greenery that can protect the settlements by reinforcing the edges from the flood. In the context of the new metropolitan epicentres, we must identify the real infrastructure caring for the elements of the territory and what elements of the past must resist. These questions open a new phase of the city that must be managed with new tools towards a new sustainability vision.

Our project proposal facing the local crisis is a strategy to promote Cultural Heritage as an agent of change, opting for a self-identity approach or settling for becoming a souvenir. First, we define which subjects as metropolitan agents can freely act within the boundaries of their metropolitan field of action (Gibson, affordances) and what are the metropolitan fields of opportunity for culture. Then, the abundance strategy is an extension of human capacities and cultures, considering Metropolitan Heritage as an 'App' fostering a valuable territory signs selection mechanism by the definition of:

(1) Generic space (bare land),
(2) Endless Combinations,
(3) Bundle of different activities to reach,
(4) Adaptability,
(5) Transformation that opens for growth.

To do so, the imprecision of the past transforms the past into an agent, the artistic grafting into borrowed words opportunity, and the different city fabrics juxtaposition into the different ways of conservation. Heritage and hetero space guide us to different metropolitan rhythms' comprehension, studying the influence of different contexts on heritage governance. The relationships between times and values are never linear and mechanical but dynamic to shape, and they are not absolute scenarios but proportions. New relationships establish among the loti and non-loti cities resonance, unseen plot possibilities and sustainable growth.

There is a functional integration of the past and vibrations (Intangible heritage), a feeling of common belonging. Heritage must trigger relevant relationships (even fragile ones) that pass through a shared value and a future project that will be realised. Visions must be projected to manage complexity: understanding reality and loving it before governing it (meta competence).

Our mapping tool allows us to define strategies towards an African city identity as a collective value and recognisable practices, which we can acquire through the regeneration of existing loti and non-loti city fabrics. Through Metropolitan Cartography, our city reading or inheritance, is diachronic. So, through the narrative infrastructure as a bridge between physical and virtual Heritage, the cultural transformation of change (the Modernity value) deals with a process of inclusion and deconstruction to read multiple identities rather than a vertical narrative against the cognitive redundancy of the mainstream governance approach.

5 Recodifying the City and Territory Model at the Metropolitan Scale in Ouagadougou

In light of the project for Ouagadougou in Burkina Faso, the narrative theme is based on the possible coexistence between the western forms of *loti* urban morphology, which resist over time, and the local *non-loti* model, which represents the extension of the city that multiplies cellularly. The hypothesis is that an extension formed by an aggregation that increases by cellular addition cannot shape the metropolitan structure that must necessarily look at a broad time horizon and be anchored to the city's chronicle. A rule of structural form that is effective concerning today's needs but that looks to the city's long spam time finds its history through the construction of the urban biography of the city.

The city fabrics' western forms that are structured in that territory can resist and modify within that resistance because the mode of structuring is characterised by a more robust capacity to resist due to the ability of the planned model to regenerate, even transforming itself. Conversely, the *non-loti* liminal spaces grow without a consensual development plan, and incompleteness allows various projects to emerge from the 'bottom' as oscillating time speed differentials in which it is difficult for anything to take hold. The extensions now pose critical questions about relationships, scale, temporalities and the incipience of differences without the necessity of physical separation. We must determine how the new metropolitan epicentres that will act as net-city nodes will be articulated with these extensions (Fig. 8).

That will only be possible if they are multipolarities in a plurality of different circulations. It is a process that reaffirms the fundamental instability of the inter-connectivity of these metropolitan territories and a potential space through which the resistance of tradition and illicit or 'informal' uses might emerge. Therefore, it requires the ability to anticipate instability and prevent disruptions by studying past data.

The form of a metropolitan city is composed of occasional and non-occasional forms. The non-occasional form of the *loti* model is characterised by an infrastructure (railways and stations + airports) that reinforces the wealth of the built settlement

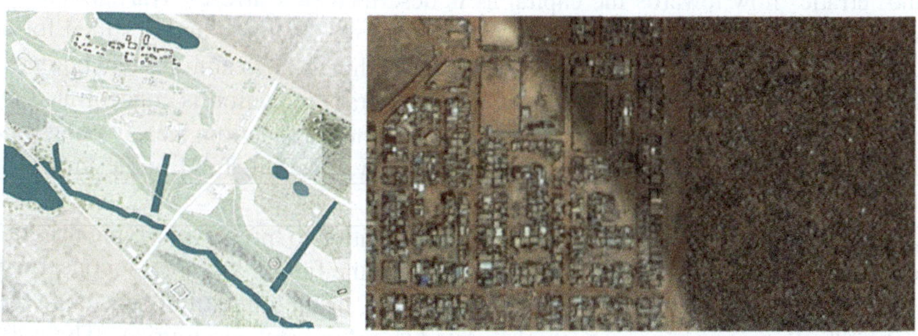

Fig. 8 Ouagadougou. Innovative campus (MSLab & Prarthana Ulaganathan, Ekaterina Kovaleva)

and infrastructure linked to the water and the canal system for the settlement and agricultural area. The public space is a qualifying element, in general, of the form of the city within schools, markets and health services that are now accepted within the new hybrid types.

The *loti* model also expresses a power of quality of existence of the city that multiplies while it does not multiply in the *non-loti* part that does not grow on itself. For the quality of the social and individual projects to increase, there must be a moment in which the city transforms and changes following the city's modified needs, bearing in mind that every form of urban settlement is a sign of established social behaviour. The *loti* model is subjected to all the diachrony of urban metabolism, meaning maintenance, renovation, and transformation must occur over time. This cycle, which we call urban metabolism, is typical of the *loti* model, which partly grows and decreases. In the ten-year circuit of its evolution, the decreased part is rebuilt. We have seen in European cities a regeneration that has been a complete transformation (Barcelona). While at first, the city grew and the centre was depopulated, at a second moment, the centre exploded, changing its function to grow in the same way which was earlier not possible.

Let us take the *non-loti* as a metropolis model. We know that the *non-loti* does not restructure itself internally since it has not understood the circuit of its urban biography, which transforms itself through urban metabolism. The *loti* has understood it; it transforms and grows through its mechanism of construction and reconstruction. So, it is necessary to determine something that can transform the *non-loti* without destroying it. Our question deals with what we should preserve of the *non-loti* and with which city growth model we should opt to determine the infrastructure that allows even the *non-loti* to preserve itself by transforming.

6 Localisation of SGDs in Ouagadougou for Design Strategies and Proposals

The challenge of the extension of *non-loti* area is the interruption of the 'nomadic' and 'erratic' flow towards the capital as is described by Carreri's Transurbanism, and the idea that the two parts of the metropolis are separated and destined for a sedentary architecture and its practices confined within de facto non-existent borders. That causes biodiversity loss because all agricultural production is concentrated in conventional food production, leaving out indigenous production. Reactivating this flow together with the idea of a circular economy linked to food is the project's theme.

The challenge of the central district is its reactivation to 'preserve' a typical architecture of the Ouagadougou metropolitan area for the future. We are not concerned with their preservation. Still, we will try to work in the interstitial spaces of the metropolis to propose services mainly related to culture and eco-tourism. The *non-loti* area's proposal concerns an urban–rural linkage pattern. A social farm pattern that can adapt to thc context in which we will insert it: urban/rural or rural/natural. Where food is produced, transformed, sold, consumed and disposed of across the boundaries of formal and informal could be the strategy and idea of the places we produce. The *loti* central one should relate more to the architectural and urban scales. However, it is also associated with the *non-loti* because, among the cultural functions, food and the last mile of the food chain could be one of the themes for the design. Here we would need the contribution of those who deal with local architecture and construction methodologies to be implemented. Also fundamental is the reading of the neighbourhood as a relational map between places. Though lacking some data, the MC activity led to what is essential to understand what data are indispensable. Our methodology should form the backbone of the analysis phases, always keeping the sequence of problem finding, setting and solving (Table 1).

Our project proposes a set of hotspots (Contin et al., 2022b) that serve four typological macro-classes: residential, productive, distributive and multifunctional. These hotspots are hybridised at emerging points, creating a new urban morphotype that can develop innovative land use, generating a linkage pattern between urban and rural areas. The functions of each hotspot define distinct spatial and organisational characteristics incorporated into the project based on its development strategy, available resources and central locations. The circularity of the idea is related to the creation of connections between all these hotspots, that will include all the resources, including information, in the network, reusing what is not used and the use of potential 'wastes' for other activities and developments.

The project features a system of low-cost housing, energy production, water purification and waste disposal that applies to all hotspots, reusing all the outputs of the different activities, to create a circular model for residential uses. This system creates buffer zones of rural–urban, natural-rural and urban-natural linkage. These buffer zones facilitate the gradual transformation and development of the agricultural periphery of Ouagadougou, turn tourism into an opportunity to implement urban services and improve the existing built environment. Moreover, the project produces a settlement pattern in the rural environment that avoids the fragmentation of family-run agricultural fields in the territory thus mitigating management problems of the same resources. The project establishes a series of pilot farms linked to a multi-service centre or agrarian consortium to achieve this vision. The initiative begins by establishing community foundations and training local leaders to enable them to develop and manage the project effectively (Table 2).

For agricultural activities, the land should be in a flat alluvial area near a river, with a water table depth of at least 10 m and a controlled average annual rainfall. In addition to constructing the necessary infrastructure, the provision of electricity and drinking water should be ensured. The land should also be hydraulically landscaped to regulate rainwater and to enable irrigation by drawing water from the river or nearby. Our project aims to solve criticalities identified with vocational courses. The cost of acquiring agricultural equipment is high due to low commercial supply, which is entirely imported. The most affordable machines are from China and India, but they are of lower quality compared to European machines (Table 3).

Table 1 Project phases

Phase	Topics addressed
Project phase one (analysis)	
1	Historical City-Territory Analysis: • Symbol, memory and image: formal references and syntactical spatial investigation of the territory's evolution • Type-morphological relationships of the historical contexts and the modifications of the type in its growth thresholds. Structure of the permanences in the qualitative change: urban biographies, genesis and geomorphology of the settlement fabrics. Gradients of formality • Identification of strategic urban areas historically suitable to support the change of scale, recombinant structures
2	Network analysis and infrastructural relation spaces: • Image and time: the potential of places in relation to territorial connections • Multi-scale relationships with multiple proximities to the structure of the territory: different nature of the networks of the territory • Interregional synergistic exchange space: large-scale interchange nodes
3	The geographical analysis of the territory: • Geo-morphological dimension of the grounds. The notion of the structure of 'support points' and definition of the operational intervention field of the project: the anthropogeographic system of the relationship between valleys (water) and neighbourhoods (grounds) • Permanence, resilience and potential of the ground structure: 'Sensitive, reactive network of proactive nodes' • Metabolic structure of the territory in relation to project possibilities of the neighbourhoods: maintenance, replacement, transformation
4	Landscape and environmental analysis: • Notations and aesthetic characters of natural and anthropic landscapes, as well as image value of local spatial relations of the territorial fabric: invariant characters and symbolic values. Iconic visual structure • Criticality, Fragility, Constraints and Transformability of the ground/land • Experimentation and mapping of landscapes: caretteriology and land use
5	Economic analysis (readings and redevelopment strategies): • Ecosystem. urban agriculture and green economy of the territory: landscape and products of the land as the essence of a lifestyle of quality • Micro and Macro Sector Economics: Analysis of the potential and economic presence on the territory (food mainly and tourism due to cultural services) that determines a possible reactivation of the territory • Analysis of the issues financed by the European Community and related formalisation of the documents necessary for participation in calls for tenders
6	Marketing analysis (local production, services and territory): • Analysis of economic and social variables: brands of the past and identification of the potential of the territory's economic network • Market analysis of potential economic realities • Identification of common structural values: towards the construction of a new brand for the Grand Ouagadougou
7	Synthesis - Situated Project: • Drafting and organisation of information: summary of the analyses of the different research sectors • Public presentation of the research
Project phase two	

(continued)

Table 1 (continued)

Phase	Topics addressed
1	Overlapping of research analysis data: • Translation of information and analysis data into summary documents: towards the definition of meta-project premises • Identification of structural permanences: construction of meta-project change • Compositional experimentations of the different operational strategies, through the forecasting of a new economic structure of the territory
2	Strategic forecasts and project map: • Formalisation of the first strategies: differentiation of visions and related operating methodologies • Definition of the economic variables of the different project forecasts • Construction of the industrial marketing potential to relaunch the district

Table 2 Actions to be activated

Action to be activated	Description
Technical and project assistance services	Collaboration with rural district technicians for agroforestry-based technical and project assistance services
Organic farming	Emphasis on crop rotations, green manures, cover crops, biocides and nematocides to promote organic farming
Multifunctional crops	Promotion of multifunctional crops for various uses, including food, pharmaceuticals, cosmetics, tools, construction, repellent and insecticidal plants, biofuels and biomass for cooking
Agreements with producer associations	Creation of a critical mass of products for buying and selling through agreements with producer associations
Cultivation contracts	Establishment of cultivation contracts with an established sale price
Supply of technical means	Provision of technical means for cultivation
Third-party machining services	Provision of third-party machining services
Accessible and sustainable agricultural credit	Provision of accessible and sustainable agricultural credit
Low-cost self-construction projects	Implementation of low-cost self-construction projects for rural buildings using local materials
Technical capacity for processing agricultural products	Development of technical capacity for processing agricultural products, such as canned tomatoes, to enable their preservation and marketing
Biomass processing for energy purposes	Processing of biomass for energy purposes for domestic use and the use of pyro-gasification stoves

Table 3 Criticalities to be solved

Criticalities Identified	Description
Difficulty in finding personnel with technical knowledge	Difficulty finding personnel with adequate technical knowledge for mechanised operations
Difficulty in finding spare parts and specialised personnel	Difficulty finding spare parts and accessories for equipment and specialised personnel for repairs
Scarcity of specialised local personnel	Scarcity of local personnel specialised in food preservation and the control of hygienic risks of food processing and occupational safety risks
Poor availability and higher costs of materials and equipment	Poor availability and higher costs than in Italy of materials and equipment useful for food processing and packaging
High certification costs for organic farming	High-quality certification costs for organic farming due to the extreme scarcity of certified farms

Regarding animal husbandry, there is a shortage of slaughterhouses that comply with European quality standards. As a result, using small container plants for raising small animals like chickens, rabbits, goats and sheep on the farm could be an alternative. The community owns the land, which can be acquired permanently or temporarily (Table 4).

The primary goal of the social initiative is to enhance the livelihoods of women and youth through the promotion of micro-entrepreneurship. To achieve this, a training program will be provided to equip them with skills for other entrepreneurial ventures such as biomass processing for domestic energy, use of pyro-gasified stoves, construction of wooden elements such as roofs and window frames, interior supplies and provision of materials for agricultural product storage, packaging and transportation. This training program will be supplemented with possible donations from various non-profit organisations. The project linked then maps to objectives. During the phases, each stakeholder in charge of the competence package tells us which data we must interpolate. They must, therefore, always be aware of why he or she produces certain maps for a project.

Table 4 Positive factors and outcomes of the project

Positive aspects
– Rapid plant growth due to favourable climate conditions
– Biodiversity offering a wide range of products to grow
– Low-cost of labour with a moderate annual increase
– Limited cost of land with the easy registration process
– Domestic market for quality agricultural products
– Network of producer associations for production and sales

7 Ouagadougou as Experimentation for CE in Non-planned Settlements

The project outcomes represent a connection between different landscapes and highlight the metropolitan dynamics that shape the transition areas. The integration of non-planned settlements into the urban management of African culture is crucial, along with the integration of scales, disciplines and governance in physical urban management. A change in the urban–rural relationship, through a formality-gradient type, will allow the non-planned settlements to benefit from the proximity to city-centre services and new mobility structures (Fig. 9).

This will also help integrate urban agriculture into the urban landscape for urban regeneration. Our study focuses on how *loti* and *non-loti* spaces and practices adapt to each other to form the boundary zone between them, which determines the *architecture of boundary objects* in this area. The purpose of our research is to present a methodological approach to describe and analyse the fundamental features of the area, establish a conceptual/physical terminology and determine the rules that govern non-planned (*non-loti*) settlements.

Nowadays, in Burkina Faso, the climate crisis is amplified by collateral effects due to interactions between local natural factors and new social models that drive the proliferation of poverty. Barrage systems that have regulated the water system since colonial times and are one of the main assets of the Burkina Capital city are not adequately maintained and can no longer fulfil their functions as ecosystem services of supply, regulation and culture (Morri and Santolini 2013). Informal settlements surround the colonial city-centre and infrastructure, resulting in a chaotic city that disrupts common urban practices, disregarding the Ouagadougou city centre's Western principle of separating harmful activities from residential areas.

Ouagadougou cannot merely be maintained or partially renovated at its current 6.9% growth rate (World Bank data). Instead, a radical transformation is necessary.

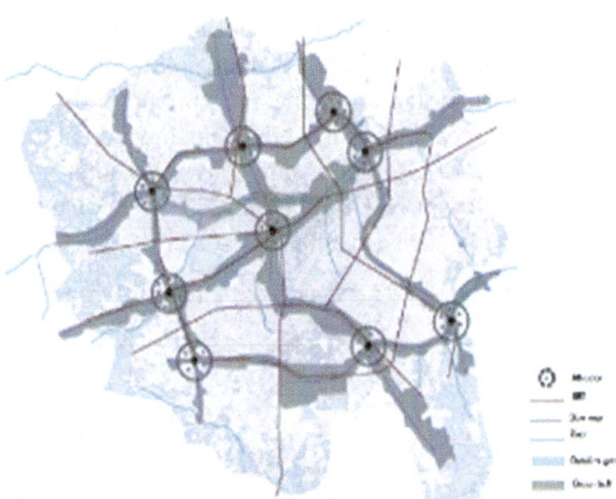

Fig. 9 Ouagadougou. Metro-Matrix

Incorporating various types of infrastructure around strategic nodes strengthens their significance and expands their impact on the surrounding area. The metropolis then becomes a patchwork of urban and suburban areas and exchange points where people gather for both environmental reasons and convenience. These exchange points contain housing, economic, and social assets that define each district. The design of the metropolis must start with a spatial concept that incorporates the strategic assets of each district into a cohesive form. Once this form is established, specific projects can be located within each district while reflecting the overall strategy.

The narrative interpreted in the light of the Ouagadougou project in Burkina Faso is based on the possible coexistence between the *loti* or planned urban morphology's western forms and the local *non-loti* city extension that multiplies cellularly. The *loti* model partly grows and decreases; the *non-loti* cannot restructure itself internally. So, it is necessary to determine which city growth model's infrastructure can transform the *non-loti* without destroying it, tracing back to the territory structure resilient points. Notable points-built maps allow studying the impacts and defining formal and informal cities and their inhabitants' lifestyles. Ouagadougou's large barrages and canals system are the infrastructures capable of forcing the public dimension as strips that absorb services and greenery, protecting the settlements by reinforcing the edges. Their long-lasting relationship with their inhabitants determines the metropolitan urban–rural linkage pattern. To improve the city circularity, we imagined a process that includes high integration levels of peri-urban farms and residential settlements, where all the outputs of the different activities can be reused as input in the following ones.

The city that serves as a model for everyone today is the smart city. So the starting question is: why is this model not admissible in Burkina Faso? To argue this, we would have to show that it is not permissible because of the cultural basis of their mental paradigm. Applying the critique of pure reason is this: knowledge of the epistemic structure of the city's entire existence can establish its construction to understand and accept its new dimension, structure and image. The actual problems to be addressed, then, are related to the food supply chain and how to determine the full cycle localisation in a metropolitan system where the refrigeration chain cannot be guaranteed everywhere, meaning that only local fresh products can be mostly sold in informal and unplanned neighbourhoods. This influences the programming of basic services, even before defining any actual localisation strategies. In the light of a new idea of CE that ensures environmental justice, ecosystem services should have an important role to play, mainly in a context where soil and natural habitats are still mostly seen as resources to consume, more than values to preserve.

References

Bruni L, Becchetti L, Zamagni S (2019) Civil economy and sustainable development. Ecra, Roma

Cohen M (2009) From master plans to stimulus packages: reflections on urban assistance to rich and poor countries by a paper submitted to 30th anniversary seminar ADP-Villes en Developpement Paris, France September 4, 2009

Conconi M, Pandolfi A (2018) Doing something scenario: teaching students an incremental approach to design strategies for development countries. IJIRES J 5(6):680–689

Contin A, Galiulo V (2022) Notes for a project mapping the accumulation of contested local intelligence in the metropolitan region. ARTdeh 9, Race

Contin A, Galiulo V, Fuentes D (2022a) Metropolitan cartography: an inventive practice tool for caring metropolitan landscape. In: Jarrett C, Sharag-Eldin A (eds) Proceedings of the ARCC-EAAE 2022 international conference architectural research centers consortium/European association for architectural education RESILIENT CITY: physical, social, and economic perspectives, Ed. Architectural Research Centers Consortium, Inc.

Contin A, Giordano P, Nacke M (ed) (2021a) Training for education, learning and leadership towards a new metropolitan discipline. Inaugural book. CIPPEC

Contin A, Ortiz P (2015) Qualitative metropolis. A new culturally encoded concept for a qualitative regeneration of formal and informal settlements. The rule and the model, TECHNE 10: Urban regeneration. FUP (Firenze University Press)

Contin A, Ortiz P, Galiulo V (2022b) The 'Unequals' inclusion. Hotspot's network strategy for a metropolitan agriculture revolution eluding informality. In: Mrinic M (ed) Informality and the City. Springer Landscape

Contin A, Ortiz PB, Della Rosa M (2013) Africa. From should to could. 49th Isocarp Congress, 2013. Proceeding

Contin A, Ortiz PB, Galiulo V, Siahkali RS, Pandolfi A, Campi P, Lutukurthi S, Sathiwada R, Kumar K, Girgaonkar P (2021b) Territorial intelligence project: governance for megalopolis urban-rural linkage pattern. Springer, Singapore. https://doi.org/10.1007/978-981-16-2023-2_3

Dantas TET, de-Souza ED, Destro IR, Hammes G, Rodriguez CMT, Soares SR (2021) How the combination of circular economy and Industry 4.0 can contribute towards achieving the sustainable development goals. In: Sustainable production and consumption, vol 26, April 2021, pp 213–227. https://doi.org/10.1016/j.spc.2020.10.005

Di Stefano A, Lepratti M (2016) Economia innovatrice. Perché è imperativo rendere circolari economia, finanza e società. ReteAmbiente, Milano

Dobbs R, Smit S, Remes J, Manyika J, Roxburgh C, Restrepo A, McKinsey Global Institute (2011) Urban world: mapping the economic power of cities. McKinsey & Company. https://www.mckinsey.com/~/media/mckinsey/featured%20insights/urbanization/urban%20world/mgi_urban_world_mapping_economic_power_of_cities_full_report.ashx

Ellen Mac Arthur Foundation (eds) (2012) Toward a circular economy. Ellen Mac Arthur Foundation, Cowes

Irajpour A, Pandolfi A (2023) The value of social integration tools: a urban perspective for the city of Piacenza. Italian J Plann Pract 12(1). ISSN 2239-267X

Korhonena J, Honkasalob A, Seppäläc J (2018) Circular Economy: the concept and its limitations. In: Ecological economics, vol 143, pp 37–46. https://doi.org/10.1016/j.ecolecon.2017.06.041

Lepratti M (2021) Circular economy. In Ramón Reyes Rodríguez Antonella Contin Patrizia Giordano, The new metropolitan voices. Glossary of metropolitan terms. Universidad de Guadalajara (Mexico)

Mayo SK, Angel S (1993) Housing: enabling markets to work. The World Bank, Washington

McKinsey Global Institute (2011) Urban world: mapping the economic power of cities, March 1, 2011, Report https://www.mckinsey.com/featured-insights/urbanization/urban-world-mapping-the-economicpower-of-cities. Press released on October 4, 2023

Morri E, Santolini R (2013) Valutare i servizi ecosistemici: un'opportunità per limitare i disturbi al paesaggio. In Battisti C, Conigliaro M, Poeta G, Teofili C (eds) Biodiversità, disturbi, minacce, Forum– Editrice Universitaria Udinese
Ortiz Castano PB (2014) The art of shaping the Metropolis. Mc Graw Hill, New York
Panchal R, Singh A, Diwan H (2021) Does circular economy performance lead to sustainable development? – A systematic literature review. J Environ Manag 293:112811. https://doi.org/10.1016/j.jenvman.2021.112811
Pandolfi A (2019) the importance of including the value of soil in metropolitan planning strategies. Int J Eng Res Technol (IJERT). ISSN 2278-0181. https://doi.org/10.17577/IJERTV8IS120359
Pagano U, Rossi MA (2009) The crash of the knowledge economy. Cambridge J Econ 33(4):665–683. JSTOR. http://www.jstor.org/stable/23601993
Rivetta A (2007) Bigness, landsmarks: la scala della regione urbana da Hilbersheimer a Koolhaas, Firenze: Alinea
Shane DG (2005) Recombinant urbanism. Conceptual modelling in architecture, urban design and city theory. John Wiley & Sons, Hoboken
World Bank (2023) Fall 2023 regional economic updates. Available here: https://www.worldbank.org/en/news/press-release/2023/10/04/world-bank-fall-2023-regional-economicupdates

Websites

https://www.worldbank.org/en/topic/urbandevelopment/overview
https://unhabitat.org/sites/default/files/documents/2019-09/strategic_plan_2020-2023.pdf
https://www.worldbank.org/en/country/burkinafaso/overview

Closing the Policy-Implementation Gaps in E-Waste Management: Implications for Circular Economy and Sustainability in Urban Ghana

Louis Kusi Frimpong, Stephen Leonard Mensah, and Seth Asare Okyere

Abstract Globally, national and city governments in developing regions are making frantic efforts to regulate and manage electrical and electronic waste (e-waste) to protect the environment and reduce its attendant environmental effects on the urban populace. Because of the above, some countries have formulated several policies and legislations aimed at tackling the problem of e-waste at different aspects of the value chain and effectively manage the negative ecological footprints often associated with e-waste proliferation. In global south cities, discussions on policies and legislations on the management of e-waste have rarely been situated in the circular economy discourse, and more so how the implementation of these policies and legislations harness circular economy principles for sustainable development. Drawing on a content analysis of scholarly literature and policy/legislative documents, this chapter aims at examining policy and implementation gaps in the management of e-waste in Ghana and its implications on the quest towards a circular economy development paradigm. We assert the imperative for e-waste management to be institutionally framed in circular economy principles to advance progress towards the sustainable urban development in Ghana.

Keywords E-waste · Policy · Institutions · Circular cities · Ghana · Sustainable development

---

L. K. Frimpong (✉)
Department of Geography and Earth Science, University of Environment and Sustainable Development, Somanya, Ghana
e-mail: lkfrimpong@uesd.edu.gh

S. L. Mensah
Department of City and Regional Planning, The University of Memphis, Memphis, TN, USA

S. A. Okyere
School of Landscape Architecture and Planning, The University of Arizona, Tucson, AZ, USA
e-mail: sethaokyere@arizona.edu

S. A. Okyere et al. (eds.), *Urban Slums and Circular Economy Synergies in the Global South*, Advances in 21st Century Human Settlements,
https://doi.org/10.1007/978-981-99-9025-2_10

1 Introduction

The transition from a predominantly rural society to an urban society has been a key feature of the development trajectory of many countries in the global south (Okyere et al, 2021). Though this structural transformation needs to be met with the needed institutional and infrastructural provisions to maximise the benefits that come with it, this has not been the case in many developing countries (Gyimah et al., 2022). Urban growth has generated significant stress for city authorities in the global south and there has been little effort to plan for the process and provide infrastructure and public services to meet the needs of most of the urban populace (Aliyu and Amadu, 2017; Shand, 2020). For instance, cities and towns undoubtedly generate more waste than rural areas (Hussein and Mansour, 2018), yet issues about how waste in all forms (solid, municipal, liquid, electronic, and electrical) can be managed in a manner that reduces the environmental health risk that comes with the poor handling of waste continue to be a challenge for city authorities in most parts of the global south (Gutberlet, 2018). This is a major sustainability issue that continues to attract attention from policymakers and researchers at the global, regional, and local levels.

One type of waste generated in urban areas, and whose management has become intractable over the last decade has been electronic and electrical waste (e-waste). E-waste refers to electrical and electronic equipment (EEE), including all its components that have reached end-of-life use or have become waste (Ongondo et al, 2011, UNEP, 2019). It includes discarded EEE such as refrigerators, washing machines, televisions, mobile phones, etc. It forms part of what is known as hazardous waste which can cause significant harm to human health if not properly managed (Igbal et al., 2017). Despite the potential harm to the environment and human health, it has generated employment for many people who are engaged in different types of activities in the e-waste value chain. This includes e-waste collectors, dismantlers, refurbishers, repairers, etc. (Oteng-Ababio et al, 2014). Indeed, it is estimated that in Ghana close to 15,000 people are engaged in refurbishing and repairs of EEE while 30,000 people are engaged in recycling e-waste (Prakash et al, 2010). The problem has always been how this activity or industry can be managed and regulated to reduce the negative impact on the environment and channel the various activities in the industry through a sustainable path.

To address the negative footprints of e-waste, there have been concerted efforts to develop policy frameworks and legislations for the management of e-waste. For instance, mention can be made of frameworks such as the Basel Convention and the Bamako Convention, with the former being a global framework and the latter a regional framework for Africa (Maes & Preston-Whyte, 2022). At the national level, there are policies and legislations which are developed based on the principles of the various global and regional frameworks but also reflecting the peculiar local challenges and aspirations. For instance, in Ghana, two legislations have been enacted to regulate EEE and e-waste use and activities. These are the Hazardous, Electronic Waste Control and Management (HEWCM) Act 2016 (Act 917) and Hazardous, Electronic and Other Wastes (Classification), Control and Management Regulations,

(HEWCMR) 2016. Aside from the policy and legal attention, there is a more important issue of a transformation in the management of waste in general, especially in such a manner that ensures that value can be derived from waste (Preston and Lehne, 2017). More so, such transformations in the management of waste are expected to tie in with recent discourse on sustainability (environment, social and economic), and a transition towards an environmentally friendly and greener waste management system. It is within this context that the circular economy (CE) has assumed much prominence in the sustainability discourse. CE is a model that allows used or waste products to be transformed into products that can be reused (Ellen MacArthur Foundation, 2016). Thus, the current discourse focuses on how institutions and policy frameworks can facilitate this waste management transformation.

The problem in most countries in the global south with regards to the management of waste, and specifically e-waste is not a matter of absence of legislation or policies, but rather how well those that have been enacted are being implemented, and whether they are being implemented in a manner that reflects principles of CE and a path towards sustainability. The above brings to the fore the persistent, yet surmountable challenges that bedevil policy implementation in Ghana, and other global south countries, some of which include institutional capacity, financing, and lack of engagement with the public. To address this issue, this chapter examines the policy/legislative implementation gaps in the management of e-waste in Ghana, and explores the missing opportunities of CE practices in the current management of e-waste activities. In doing so, the chapter addresses the following objectives:

1. Examine current policies and legislations on e-waste management in Ghana
2. Examine informal and formal e-waste management practices in Ghana
3. Explore the gaps and missing links between policies and actual practices and their implications towards a CE model.

The chapter is structured as follows. After the introduction the chapter discusses the CE model, focusing on its meaning and experiences with its application in developing countries. The next review is on the policy-implementation gaps. The methodology is next presented. After the methodology is a review of policies and legislations on e-waste management in Ghana and an empirical literature on e-waste management practices in the formal and informal sector in Ghana. Following this is a discussion on the missing links between the legislations and practices on the ground, and the implications this may have on our quest towards a CE model. The last section in the conclusion and recommendations.

2 Circular Economy and the Global South Policy Challenges

The circular economy (CE) is a model that is in line with the sustainable development agenda as it aims to protect and conserve the environment through innovative integration of natural resources, business, and efficient waste management strategies (Ogunmakinde et al., 2021). It stands in contrast to the linear economy, which comprises the process of taking natural resources, turning them into products, and then discarding them once they have been consumed. The circular economy aims to close this gap through the adoption of innovation, where used products can get back into the production cycle (Wang et al, 2022). By so doing, it aims to eliminate or reduce waste through the reuse and recycling of materials that have reached their end-of-life stage. The CE model has become the preferred business model for most industries as its principles align with the shift towards clean and renewable energy and environmental sustainability. According to OECD (2021) the circular economy aims to redefine economic growth by designing waste management systems that allow products to be in circulatory thereby leading to the restoration and regeneration of products.

The principles underlying CE have evolved over the years, though for most studies on the subject, the 'R' alphabet principles have been maintained (Hu et al., 2011; Hartley et al, 2023). Prior to 2011, discussions on the principles of CE centred on the 3Rs, which include reduce, reuse, and recycle. Reuse refers to the use of a product more than once. Recycling involves the reintroduction of waste materials into the production stream as raw materials so that they are transformed into new products. Reduce on the other hand involves a decrease in the use of raw materials and the use of alternatives with the aim of protecting the environment. Thus, responsibilities are placed on both consumers and industries when it comes to 'reduce'. Recent studies on the principles have expanded the Rs, indicating that the list can be inexhaustive. For instance, Reike et al (2018) came up with 38R CE principles, while Mhatre et al (2021) have also come up with 45R CE principles. The expansion of the CE principles indicates the broadening of the range of sectors that they can be applied to, and how they can be adapted to meet the strategic standpoint of different enterprises (Ferasso et al., 2020). More recently, Hartley et al, (2023) proposed a list of 60R CE principles that were subsumed under a broader 4R CE principles which were R1 (reuse), R2 (reduce), R3 (recycle), and R4 (reverse logistics).

Aside from the principles which are being integrated into the business models of companies worldwide, discussion on CE has also centred on CE transition through institutional and policy support (Hartley et al, 2023). For instance, policy frameworks in several countries are moving away from spot-level initiatives towards frameworks that enhance sustainability and circularity which is reflected in interventions such as regulation of operations, investment opportunities, sustainable environmental practices, and institutional strengthening (Antwi-Afari et al., 2022). Nonetheless, it has been suggested that despite the varied interventions needed in sectors such as waste management, a more comprehensive policy instrument/framework can be more transformational, albeit, it should be accompanied by measures that can enhance the practicability in their implementation. Hartley et al (2023) argue that policy frameworks on CE should take into account the life cycle of materials, and how they are handled at every stage of their life cycle to ensure that the social, economic, and environmental impacts are minimal.

Although integrating CE into policy is necessary for a sustainable waste management and urban development in developing countries, its implementation faces significant obstacles. The adoption of CE is hampered by three primary issues, according to Preston et al. (2019). These issues are capacity and financial limitations, economic and structural problems, and lack of infrastructure and urbanisation. The terms "capacity and finance constraints" combine two distinct problems, namely institutional capacity and access to finance limits. Institutional capacity refers to the governance framework, which includes the institutional structures needed to coordinate and implement these policies and regulations. The situation in most developing countries is characterised by weak regulatory and enforcement regimes, which often stem from inefficient bureaucracy and ineffective institutional coordination (Adekunle, 2021; Windapo & Rotimi, 2012). For access to finance, it has been argued that the shift towards CE requires investment in infrastructure, industrial development, innovation, and technological access (Preston et al., 2019). Most developing countries do not have such finances and often depend on foreign direct investment (FDI) or concessional financing to be able to make such investments. Even at the institutional level, adequate finance is required to monitor and regulate different streams of activities necessary for CE and this is often not the case in most developing countries.

The second challenge i.e., economic and structural encapsulates the informal nature of the economy in most developing countries. This refers to the unorganised nature of the private businesses, especially those engaged in waste management. The problem is that data on various activities are unavailable, which also affect the quality of information needed for planning, monitoring, and revenue generation (Navarrete-Hernandez & Navarrete-Hernandez, 2017). Further, the larger informal sector also means that a large proportion of people engaged in waste management also lack the skills and technology needed to optimise waste collection and recycling. The third challenge which is infrastructure deficit and urbanisation has to do with the constraints of having access to critical infrastructure needed to facilitate industrial development and a system that can enhance the circulation of waste products. More so, there are far smaller stocks of materials for reuse and recycling (International Resources Panel, 2012).

In view of the relevance of policy formulation, and more importantly how their successful implementation can be crucial to the transitioning to a CE, the next section discusses the policy formulation process, and factors that accounts for gaps in the implementation of policy frameworks.

3 Policy-Implementation Gaps: An Overview

Policymaking is a process that involves putting together decisions, plans, and policies to guide a system's operations and governance (Abdullahi and Othman, 2020; Cerna, 2013). The policy lifecycle has two major facets: policy formulation and policy implementation. Policy formulation involves addressing a problem by creating a policy document or framework. Policy implementation involves translating ideas into actions, with diverse organizations, stakeholders, experts, and administrators working together (Hill and Hupe, 2015). In developing countries, policies are rarely carried out, leading to a wide gap between policy intent and implementation (Mugwagwa et al, 2015).

According to Abdullahi and Othman (2020), studies on policy-implementation gaps began in the 1970s and since then, there have been three different generations of studies in this area. The first-generation studies which were between 1970 and 1980 focused on policy-implementation failures arising out of a top-down governance approach. These studies showed that lack of community buy-in, lack of inclusiveness, and poor understanding of policies by other stakeholders at the lower level often derailed the implementation of these policies (May 2014). The second-generation research was in response to the first-generation and emphasised bottom-up approaches to policy implementation. It focused on issues about capacity building for local implementers of policies, lower-level bureaucracy, and engagement with stakeholders. The third generation of studies on policy and implementation research which began in the late 1980s focused on building synergies between central and local-level bureaucracy and developing frameworks that reconcile top-down and bottom-up approaches. Here, the emphasis was on how policies developed from the top with local-level involvement can be implemented at the local level in such a manner that will minimise tension and reduce implementation failures (Cairney et al., 2013).

Hudson et al. (2019) identify four main factors contributing to policy-implementation gaps: overly optimistic expectations, implementation in dispersed governance, inadequate collaborative policymaking, and the vagaries of the political cycle. Optimistic expectations result from underestimating the problems at hand, leading to underestimation of the cost, time, and resources required for policy implementation. Implementation in dispersed governance requires the involvement of sub-national governing structures and stakeholders, and effective coordination between central and sub-national levels for successful policy implementation. Inadequate collaborative policymaking results in poor engagement and collaboration among government institutions and stakeholders outside the government apparatus (Gazley, 2017). The vagaries of the political cycle can lead to short-term policies that do not address problems in their entirety, compromising long-term policy decisions.

To address policy-implementation gaps, Hudson et al (2019) propose the following measures to address the problem. These are policy tracking, implementation support, and implementation review. Policy tracking involves the establishment of a unit, entity, or structure to track the progress of the implementation of policies. Gold

(2017) suggests that these measures ought to be taken to ensure effective policy tracking. These measures are performance monitoring, problem-solving activities, and progress assessment. Implementation support on the other hand refers to the support mechanisms needed to support the implementation of policies. At the centre of this are the implementing agencies of the policy. The support can come in two ways. These are the support the implementing agency ought to receive from stakeholders, and the other is the support the agency ought to give to other stakeholders whose roles are critical to the successful implementation of the policy. Some implementation support highlighted by Hudson et al (2019) includes capacity building, problem-solving solutions (technologies), financial, management, and regulation. The last which is implementation review refers to the opportunity of assessing policy objectives, focus, and actions. Implementation review allows policymakers and implementers to modify trajectories and implementation support mechanisms.

4 Methodology

The methodological approach for the chapter comprised a content analysis of policy and legislative documents on e-waste in Ghana. There was also a review of literature in two separate sections of the chapter. First was the review of literature on CE policy challenge experiences in the global south and literature on policy-implementation gaps. The above literature reviewed was to provide the context for discussing the policy-implementation gaps with regards to e-waste in Ghana. The second section of the review, which was part of the result was on the empirical literature on e-waste management and practices in Ghana. The content analysis of policy documents was useful because it allowed for the search and analysis of the presence and meaning of text, themes, and concepts (Heath and Swabey, 2014). Still on the policy and legislative documents, the focus was on key texts relating to the handling and management of EEE and e-waste which include disposal, collection, recycling, reuse, refurbishment, etc. (i.e., management and circularity of e-waste), the responsibilities of stakeholders regarding the handling and management of e-waste, and measures to ensure the circularity and sustainable management of e-waste. The content analysis was conducted on three main documents which include National Environmental Policy, Hazardous and Electronic Waste Control and Management (HEWCM) Act 2016 (Act 917) and Hazardous, Electronic and Other Wastes (Classification), Control and Management Regulations, (HEWCMR) 2016. The HEWCM Act 2016 and HEWCMR Act 2016 were selected for the content analysis because they provide the framework for e-waste management and handling in Ghana. The NEP was included because it is the overarching policy document for environmental management in Ghana, and e-waste is one aspect of environmental problems that require attention in the country, hence it was important to analyse what the policy provides regarding e-waste. Other environmental policies were not included and these were Ghana Landfill Guidelines (2002) and Manual for the preparation of district waste management plans in Ghana (2002).

This is because these policy documents do not specifically focus on the management of e-waste.

Regarding the empirical review, research and scholarly on e-waste management and practices in Ghana, articles from media sources, and published reports were reviewed. The research and scholarly articles were sourced from Google scholar and the Scopus database. The search process in the two data bases were guided by key phrases such as 'e-waste practices in Ghana', 'e-waste management in Ghana', 'e-waste recycling in Ghana', and 'institutional arrangement for implementing policies and regulations on e-waste in Ghana'. Scholarly articles retrieved from the search engines were screened and their appropriateness for the review were assessed. About 20 scholarly articles, reports, and news articles were used for the empirical review of e-waste practices in Ghana.

5 Review of Legislative and Policy Frameworks

5.1 National Environmental Policy

The National Environmental Policy (NEP) is the country's foremost policy on the sustainable management of the environment. The goal of NEP is to ensure citizens' quality of life and their living and working environments through sustainable development based on integrated and coordinated environmental management. The first environmental policy formulated in 1994 suggested a restructuring of the lead agency on environmental matters in the country (the Environmental Protection Agency (EPA) to spearhead the drive towards sustainable environmental issues in the country. In terms of e-waste management, there are provisions on the characteristics and interventions of e-waste. The characteristics of e-waste as stated in the policy document are the sources: 'used equipment in the form of computers, copying machines, television sets, mobile phones, and electronic equipment' (Government of Ghana, 2000 p 6); it is rejected in the country of origin and imported into the country without regard to age and degree of usefulness; no guideline on disposal technique, no e-waste or recycling programme, burning of e-waste to retrieve useful parts and the detrimental impact of emissions on human life. In terms of the interventions, the provision in the policy document is more general and refers to the EPA Act 1994 which provides the framework 'for waste management through the principles that refer to avoidance or minimization and remediation of pollution, including waste reduction reuse, recycling, and proper waste disposal' (Government of Ghana, 2000 p.7). Thus, this principle that reference is made to, applies to all forms of waste and not specifically to e-waste.

The NEP, as can be seen, has a broad focus in that it seeks to address all forms of environmental challenges facing the country and, more so, provide information on specific regulations and institutions mandated to handle these environmental challenges. Because of this, the policy lacks details on how e-waste should be managed.

Reference is made to the EPA Act 490 on the interventions (Government of Ghana, 2000, p 7). However, the EPA Act 490 provides the framework for administrative procedures for managing waste in general and not specifically on e-waste.

5.2 *Hazardous and Electronic Waste Control and Management (HEWCM) Act 2016 (Act 917)*

This legislation outlines the control, management, and disposal of hazardous waste, electrical and electronic waste, and related purposes. The Act is segmented into two main parts. Part one deals with hazardous waste and other wastes, while part two of the Act deals exclusively with electrical and electronic waste. The main sections under part two include electrical and electronic waste levy, electrical and electronic waste fund, and electrical and electronic waste recycling and other provisions (Government of Ghana, 2016a). With regards to the levy, the Act states that the import of specific electrical and electronic items (see Government of Ghana, 2016a p.32–40) shall attract the payment of a levy termed 'eco levy'. It also states that all electrical and electronic waste imported into or exported from the country will have to be done after a permit is sort from the mandated or authorised Agency. The Act also states that 'A retailer, distributor, or wholesaler who is into manufacturing of electrical and electronic waste shall register with the Agency particulars of electronic and electrical equipment manufactured or assembled in the country' (Government of Ghana, 2016a p 15). The Act further mentions the establishment of an 'Electronic and Electrical Waste Management Fund'. Among the objectives of the fund include; the construction and maintenance of an e-waste recycling or treatment plant; research into methods of e-waste preservation, prevention, and control; research into e-waste recycling; offer of incentives for collection, transportation, and disposal of e-waste; monitoring, compliance, and enforcement to mention but a few. The main source of money for the fund includes the eco levy, monies approved by parliament, other monies received from other sources and lawfully payable to the fund, which includes grants, donations, and gifts (Government of Ghana, 2016a).

The Act mentions the establishment of a recycling plant as stated, '*The Minister, on the advice of the Agency may make arrangements for the establishment of electrical and electronic waste recycling plant and related facilities in the country*' (Government of Ghana, 2016a, p 18). The Act further mentions that the manufacturer, distributor, or wholesaler of electrical or electronic equipment '*shall take back the used or discarded electrical and electronic equipment manufactured for recycling purposes*' (Government of Ghana, 2016a, p.18), while persons responsible for recycling and disposal shall ensure that it is done in an environmentally sound manner. The Act mentions that the Agency shall ensure that the above-mentioned provisions on the disposal and recycling of electronic and electrical equipment shall be carried out and will also take steps to apply the necessary sanctions and penalties if the provisions are not adhered to.

The HEWCM, Act 2016 is one of the first regulatory frameworks to offer comprehensive detail on the management of e-waste in Ghana. It touches on almost all areas of the management of e-waste including financing, sanction, and regulations of activities of all actors in the value chain. Nonetheless, there are several challenges or weaknesses that can be identified with the HEWCM, Act 2016. First, though the eco levy is an important source of financing e-waste management in the country, the Act in its current state is not clear on the timelines for the disbursement of the funds. Further, though it is stated that the 'Electronic and Electrical Waste Management Fund' will also be used to finance some activities such as the establishment of a recycling plant, and funding research activities on e-waste control, and recycling, there is no timeline, and procedures for tracking these activities to be used by the fund is also not clearly stated. Additionally, the proportion of the funds that should go into these activities is also unclear. Further, it is assumed that all persons engaged in various activities in the e-waste value chain will be able to take back or reclaim discarded electrical and electronic waste, however, this borders much on the capacity of persons so involved. This is where the regulation also fails to provide support mechanisms or procedures to support the informal sector to enable persons in the sector to take back and properly account for e-waste stored and recycled.

5.3 Hazardous, Electronic and Other Wastes (Classification), Control and Management Regulations, (HEWCMR) 2016

The HEWCMR is an Act of Parliament that aims to provide guidelines on how the different classifications of hazardous and electronic waste must be treated, handled, and managed, including the roles that must be played by different stakeholders in this process. Two sections are dedicated to e-waste and include the sections on 'Electrical and Electronic Equipment Waste' and 'Treatment of Electrical and Electronic Waste'. The former section outlines the responsibilities of various stakeholders in the e-waste management stream. For instance, the Act states that the producer or manufacturer of EEE is responsible for the following; 'ensure the collection of e-waste that it generates; ensure the recycling and disposal of e-waste that it generates during manufacturing; collect the e-waste that the producer generates from the products in accordance with the Extended Producer Responsibility; set-up a track-back system; must have adequate finance system to promote environmentally best practice' (Government of Ghana, 2016b p.39). The refurbisher is also required to properly store the e-waste that it collects; ensure that e-waste is recycled or disposed of properly after refurbishing; ensure proper record keeping of e-waste safe the refurbisher has handled (Government of Ghana, 2016b). Like the above categories of e-waste workers, collectors are required to ensure safe and secure storage of e-waste until it is sent to the refurbisher; ensure that damage is not caused to the environment while storing and transporting e-waste; and maintain proper records of e-waste. For the recyclers, their responsibilities include to; ensure that recycling facilities are established in accordance with the guideline provided by the Agency; making available to the Agency the necessary records regarding their operations; and ensure that waste arising out of the recycling process are disposed of safely. More so, all the above-mentioned categories of e-waste processing must obtain the necessary permit from the supervisory Agency which is the EPA.

In terms of the treatment of e-waste, the Act mentions that producers or manufacturers of EEE shall maintain records on the collection, sale, transfer, storage, and separation of EEE and make them available to the Agency. The Act also states that disposal of e-waste is not permitted in any waste facility in any District Assembly unless the disposal is in accordance with standards stated by the Agency. The Act states that all transporters of EEE shall register with the Agency and must obtain the necessary permit to carry out their activities. The same also applies to persons or organisations that are into the treatment of EEE. Lastly, the Act also mentions that there will be a system in place to ensure the recovery of e-waste. Details in terms of the nature and structure of this system are not clear in the Act. Further, it also states that producers or manufacturers shall put in place measures to meet the following targets; (1) recover 85% of specific types of EEE (2) reuse and recycle 75% of specific types of EEE (see Government of Ghana, 2016b p. 148).

The HEWCMR, Act 2016 provides details regarding the various actors within the e-waste value chain and also has clear provisions on the responsibilities of these identified actors in the management of e-waste. Further, Act also provides guidelines on how e-waste must be handled by the various actors within the chain. Nonetheless, there are practical challenges that can hinder its implementation on the ground. First, the emphasis on record keeping, storage systems, an appropriate track-back system, and a finance system to promote environmental best practices are all important steps towards the formalisation of the e-waste industry, and to enhance better monitoring by the Agency. Unfortunately, with the sector largely dominated by the informal sector, which is underfinanced, under-skilled, and rarely supported, it remains to be seen how these measures can be adequately adhered to by actors in the informal sector. Aside from the above, the Act also has no clear provision on measures that the Agency will take to ensure that the set targets for recovery, reuse, and recycling is achieved.

6 Review of Empirical Literature on E-Waste Practices in Ghana

E-waste management practices in Ghana are largely dominated by the activities of the informal sector. The sector is comprised of different categories of workers which include scrap dealers, waste scavengers, waste pickers, dismantlers, refurbishes, and open dealers. Due to the lack of appropriate technology, workers in the informal sector predominantly utilise crude methods in their operations. These include open burning, open dumping of e-waste at landfill sites, and the repair, reuse, or mining of valuable resources (Owusu-Twum et al., 2022). Agbogbloshie, a slum community in Accra is an internationally known centre for e-waste processing (Grant & Oteng-Ababio, 2012). The way of life of residents is intertwined with e-waste activities, which creates a source of livelihood and valuable resources for their survival. Materials from e-waste such as scrap metals are also recycled for designing new materials for consumption (Daum et al., 2017). The Agbogbloshie site provides livelihood opportunities to about 6000 workers and additional 1500 persons indirectly (Grant & Oteng-Ababio, 2012). The lack of appropriate technologies and safety guidelines in e-waste recycling poses health risk to families and threatens ecological sustainability (Daum et al., 2017). For instance, anecdotal evidence suggests that airborne and other harmful chemicals have infiltrated the body tissues of residents and causing massive environmental pollution (Daum et al., 2017; UNEP, 2013; Amfo-Otu et al., 2013).

Integrating the activities of the formal sector in e-waste management practices is crucial for realising equitable opportunities for all stakeholders. Yet, a paltry 1% of generated e-wastes is recycled by the formal sector, while the remaining is recycled by the informal sector (UNEP, 2013). The increase in e-waste generation is the result of rising digitalisation initiatives and the use of digital technologies. While these have benefitted many countries especially in the Global North, huge volume of generated waste and related problems have been the visual outcome in developing countries particularly Ghana (Manmohit et al., 2018; Ohajinwa et al., 2019; Amfo-Out et al., 2013). Imported EEE are done by formal businesses and manufacturers and are brought into the country in the form of charitable and second-hand goods (Amoyaw-Osei et al. 2011). Activities of informal importers also contribute to the volume of generated e-waste as the quality of imported goods decreases with the engagement of informal importers (Daum et al., 2017). Informal importers through their actions of bringing into the country used digital products have been noted to significantly contribute to e-waste generation (Grant & Oteng-Ababio, 2016).

The Government of Ghana has made efforts to clamp down on e-waste importation and generation. For instance, in 2016 the Hazardous and Electronic Waste Control and Management Regulations (Act 917) was enacted to require producers and importers to register with the EPA for the payment of taxes and levies (Owusu-Twum et al., 2022). The government has also engaged the services of the private sector to manage and recycle about 40% of the generated e-waste (Akon-Yamga et al., 2021). Considering that e-waste generation causes significant environmental pollution and health risk across the globe, its management requires adequate policies and strategies to safeguard human health and the integrity of the environment (Kumar, 2017). Collaborative and equitable planning responses targeted at e-waste management have numerous socioeconomic, environmental, and health benefits (Cucchiella et al. 2015). Managing e-waste in Ghana has become very challenging as most policies designed to control the import and handling of e-waste marginalise the informal sector and emphasise the maximisation of economic benefits such as accruing revenues from levies and taxes (Akon-Yamga et al., 2021). Due to this, little strides have been made to improve activities in the sector since many informal sector players continue to use rudimentary practices and technologies to handle and dispose e-waste.

The efforts of government towards the sustainable management of e-waste have also been inefficient due to policy incoherence, inadequate funding, and weak collaboration between formal and informal actors. Legislative and policy responses do little to include sociocultural and environmental attributes of residents which contribute to e-waste generation. Further, institutions mandated with oversight on environmental regulations are present in a few districts and regions and constrained by logistics and finances. For instance, the EPA currently has 10 out of 16 regional offices and lacks a presence in many district offices (EPA, 2021). Behavioural challenges resulting from resident disregard for environmental regulations have also partly contributed to the current e-waste menace. This is evident in many slum communities where digital products are openly burnt or thrown into open spaces even though such practices are detrimental to the health of residents and the sustainability of the environment The future of e-waste in Ghana's circular economy requires massive public education, adequate financing, resourcing institutions and the provision of alternative livelihoods for workers in the informal sector.

7 Policy-Implementation Gaps in E-waste Management

The reality of e-waste management and practices indicates that there are gaps in the implementation of the legislation, and this is also evidenced by the challenges that are encountered, especially by the implementing agency i.e., EPA. For instance, the HEWCMR (2016) provides specific provisions on the handling of e-waste within the various streams of activities or categories of workers. This includes safe storage, safe handling, and record keeping. Unfortunately, this has not been the case. This can partly be attributed to the informal nature of most of these activities, where workers have no adequate training and knowledge of the safe handling of e-waste (Navarrete-Hernandez & Navarrete-Hernandez, 2017). Further, safe handling also requires access to safe technologies which is a challenge for most informal e-waste workers. On the other hand, the chapter also shows a failure on the part of the implementing agency to monitor the informal e-waste activities and engage with people in the sector to support them through training and access to technologies. This point ties in with Hudson et al. (2019) argument that inadequate collaboration and engagement can derail or affect the smooth implementation of policies. Further, it also ties in with Fobbe and Hilletofth (2023) argument that policies that aim to facilitate CE transition must prioritise stakeholder engagement. Indeed, a similar opinion is expressed by OECD (2020) which argues that CE transition is a shared responsibility across levels of government and stakeholders and that the successful implementation of CE policies hinges on how well institutions and stakeholders can define their responsibilities and commit to working together to attain a common goal.

In addition to the above, the legislation on e-waste places emphasises on recycling of e-waste and the various activities necessary to facilitate e-waste recycling. This includes the establishment of a levy to support the setting up of recycling facilities and the collection and transportation of e-waste. The reality, however, is that a substantially small amount of e-waste undergoes recycling (UNEP, 2013) 'i.e., transformation into new products'. Ideally, the formal sector which comprises manufacturing industries involved in the production of EEE and those in e-waste recycling are best to handle these recycling activities. Unfortunately, the number of formal industries in Ghana is very few, and only recycles just 1% of e-waste products in the country (UNEP, 2013). On the other hand, recycling is rarely done in the informal sector. Rather, the informal sector is mostly engaged in the collection, dismantling, and retrieval of valuable metals which are sent to some recycling facilities, which then export most of them (Grant and Oteng-Ababio., 2012). The problem, however, is that the informal sector by its nature and modus operandi is unable to collect and retrieve a large amount of e-waste (Oteng-Ababio et al., 2014). Aside from the inadequate recycling facility, there is the challenge of developing an efficient system that can support the recycling of e-waste. Indeed, there are examples of policy and legislative support for recycling in other countries that can be learnt from. For instance, in Denmark, a national subsidy scheme has been instituted to support industries engaged in wastewater. Similarly, the CE Investment Fund in Scotland also offers SMEs GBP 18 million in grant money each year to help them make their business models more circular (OECD, 2020).

Further, implementation failure of the various legislations on e-waste stems from systemic failures in the implementation support that ought to be given to EPA as an implementing agency. For instance, the EPA is required to research e-waste recycling or partner with other bodies to undertake such research activities. The EPA is also required to prevent and control, monitor, and ensure compliance with the regulations on e-waste. These activities however require sufficient funding which happens to be the main nemesis of the agency. Though the eco levy was set up to fund some of these activities of the Agency, the implementation of the eco levy has also not been without challenges. For instance, though the collection of the fund began in 2018, the fund is yet to be disbursed (Ghana Business News, March 2022). Aside from this, EPA's presence in the country is also low. At present, the EPA has 10 regional offices out of the 16 regions in Ghana and no offices at the district level.

The transition towards CE requires deeper reflections in policy and legislative instruments that move beyond the business-as-usual policy frameworks that tackle the problem at one end and leave others at the other end. It is important that legislative and policy frameworks on CE get the buy-in of players in the industry, backed by government commitment through policy and institutional reforms, financial commitments, and active engagement with players in the industry, especially those in the informal sector (Preston et al., 2019). This is critical in closing the policy implementation gaps that have been identified in this chapter. Though the review of the legislative documents revealed provisions aimed at facilitating storage, recycling, and segregation of e-waste through adherence to guidelines by the Agency (see Government of Ghana, 2016b pg. 39), there must be concerted efforts to support the industry players to invest in recycling, increase the competitiveness of industry players, while ensuring strict compliance to environmental standards. The latter requires institutional strengthening to improve monitoring, compliance, and better coordination with other cognate government institutions and industry players.

8 Concluding Remarks

The management of e-waste continues to present a challenge to authorities in the global south. To help address some of these challenges, efforts have been made to enact several legislative instruments as a means of providing some guidelines for its management. The purpose of this chapter was to examine the policy/legislative-implementation gaps in the management of e-waste in Ghana and explore the missing opportunities of CE practices in the current management of e-waste. The study sought to provide useful insights into areas that need attention in the implementation of legislations on e-waste and highlight the inefficiencies in the current practices and its implications in attaining CE and a path towards sustainability.

The findings show that there are two main legislative documents regulating e-waste activities in Ghana, and these are the Hazardous and Electronic Waste Control and Management (HEWCM) Act 2016 (Act 917) and Hazardous, Electronic and Other Wastes (Classification), Control and Management Regulations, (HEWCMR) 2016. The former has provisions on the eco levy, what the levy will be used for and the registration of activities within the e-waste value chain. It also mentions how discarded EEE shall be taken back for recycling purposes. The latter on the other hand outlines the responsibilities of various stakeholders in the e-waste management stream. It also highlighted the treatment of e-waste and the responsibilities of manufacturers in the collection, sale, transfer, storage, and separation of EEE. The chapter revealed that activities on the ground show that e-waste is not handled and managed in a manner that ensures safety to workers (open burning of waste metals), while the sector is largely dominated by the informal sector which has little capacity to recover and recycle e-waste. Further, the formal sector with the capacity to recycle, is relatively few, with most of them rather engaged in the import of EEE, with very few engaged in e-waste recycling. Several gaps were identified in the implementation of the regulation on e-waste, with many of these gaps attributable to implementation challenges on the part of the EPA.

The chapter is not without some limitations. First, the chapter was limited to the review of policy and legislative documents, as well as empirical literature. Opinions and perspectives of actors in the sector were not included. Thus, we were limited in benefitting from triangulating data from different sources. Second, a comparative study of policy and legislative documents with other countries in the sub-region would have provided very good perspectives on what other countries are doing and how we can learn from them to improve the situation in Ghana. However, such a study can be looked into in other future studies on CE policy implementation in Ghana. This said the chapter offers a few recommendations that should be taken on board. First, support must be given to the implementing agency so that it can successfully carry out its task of regulating the e-waste sector. The support must include financial support and increase the capacity of the EPA. In this regard, we entreat the administrator of the eco fund to begin with the allocation of the fund, while EPA must also do well to establish offices through the remaining six regions. Further, we also suggest that EPA must engage with other stakeholders especially those in the informal sector. This can facilitate the registration of informal businesses, which will enhance the monitoring and tracking of their activities. Through this, workers in the informal sector can be trained on safe and efficient ways of recycling, storage, and retrieval of e-waste materials. Lastly, we recommend that systems must be put in place to promote linkages between the formal and informal industries engaged in e-waste activities. This can enhance employment opportunities for both sectors and increase production for both sectors.

References

Abdullahi M, Othman N (2020) Bridging the gap between policy intent and implementation. J Sci, Technol Innov Policy 6(1):1–10

Adekunle IA (2021) On the search for environmental sustainability in Africa: the role of governance. Environ Sci Pollut Res 28:14607–14620. https://doi.org/10.1007/s11356-020-11432-5

Akon-Yamga G, Daniels CU, Quaye W, Ting BM, Asante AA (2021) Transformative innovation policy approach to e-waste management in Ghana: Perspectives of actors on transformative changes. Sci Public Policy 48(3):387–397

Aliyu AA, Amadu L (2017) Urbanization, cities, and health: the challenges to Nigeria-a review. Ann Afr Med 16(4):149–158. https://doi.org/10.4103/aam.aam_1_17

Amfo-Out R, Bentum JK, Omari S (2013) Assessment of soil contamination through e-waste recycling activities in Tema community one. Environ Pollut 2:66

Amoyaw-Osei Y, Agyekum OO, Pwamang JA, Mueller E, Fasko R, Schluep M (2011) Ghana e-waste country assessment: SBC e-Waste Africa Project. https://www.basel.int/Portals/4/Basel%20Convention/docs/eWaste/E-wasteAssessmentGhana.pdf

Antwi-Afari P, Ng ST, Chen J (2022) Developing an integrative method and design guidelines for achieving systemic circularity in the construction industry. J Clean Prod 354:131752

Cairney P, Nilsen P, Ståhl C, Roback K (2013) Never the Twain shall meet?-a comparison of implementation science and policy implementation research. Implement Sci 8(63):1–12

Cerna L (2013) The nature of policy change and implementation: a review of different theoretical approaches. https://www.oecd.org/education/ceri/The%20Nature%20of%20Policy%20Change%20and%20Implementation.pdf

Cucchiella F, D'Adamo, I, Koh SCL, Rosa P (2015) Recycling of WEEEs: an economic assessment of present and future E-waste streams. Renew Sustain Energy Rev 51:263–272

Daum K, Stoler J, Grant RJ (2017) Toward a more sustainable trajectory for e-waste policy: a review of a decade of e-waste research in Accra, Ghana. Int J Environ Res Public Health 14(2):135

Ellen MacArthur Foundation (2016) Circular economy in India: rethinking growth for long-term prosperity. https://www.ellenmacarthurfoundation.org/assets/downloads/publications/Circular-economy-in-India_5-Dec_2016.pdf

Environmental Protection Agency (2021) History of EPA. https://epaoilandgas.org/history%20Of%20EPA/index.html#:~:text=EPA%20Regional%20Offices,10%20administrative%20regions%20of%20Ghana

Ferasso M, Beliaeva T, Kraus S, Clauss T, Ribeiro-Soriano D (2020) Circular economy business models: the state of research and avenues ahead. Bus Strateg Environ 29:3006–3024. https://doi.org/10.1002/bse.2554
Fobbe L, Hilletofth P (2023) Moving toward a circular economy in manufacturing organizations: the role of circular stakeholder engagement practices. Int J Logist Manag 34(3):674–698. https://doi.org/10.1108/IJLM-03-2022-0143
Gazley B (2017) The current state of inter-organizational collaboration: lessons for human service research and management. Hum Serv Organ: Manag, LeadShip & Gov 41(1):1–5. https://doi.org/10.1080/23303131.2015.1095582
Ghana Business News (2022) E-waste law: Ghana government to collect eco-levy on additional 300 electric products. https://www.ghanabusinessnews.com/2022/03/19/e-waste-law-ghana-government-to-collect-eco-levy-on-additional-300-electric-products/#:~:text=300%20electric%20products-,E-waste%20law%3A%20Ghana%20government%20to%20collect%20eco-levy,on%20additional%20300%20electric%20products&text=Government%20through%20the%20Ghana%20Revenue,boost%20the%20country%27s%20revenue%20generation
Gold J (2017) Tracking delivery global trends and warning signs in delivery units. Institute for Government, London
Government of Ghana (2000) National environmental policy of Ghana. https://new-ndpc-static1.s3.amazonaws.com/CACHES/PUBLICATIONS/2016/04/16/NATIONAL_ENVIRONMENTAL_POLICY.pdf
Government of Ghana (2016a) Hazardous and electronic waste control and management (HEWCM) Act 2016 (Act 917). https://www.epa.gov.gh/epa/sites/default/files/downloads/publications/Hazardous%20and%20Electronic%20Waste%20Control%20and%20Mgt%20Act%20917.pdf
Government of Ghana (2016b) Hazardous, electronic and other wastes (classification), control and management regulations, (HEWCMR) 2016. http://ir.parliament.gh/bitstream/handle/123456789/383/FINAL%20HAZARDOUS%20AND%20ELECTRONIC%20WASTE%20CONTROL%20AND%20MANAGEMENT%20%20BILL,%202016.pdf?sequence=1
Grant R, Oteng-Ababio M (2012) Mapping the invisible and real African economy: urban E-waste circuitry. Urban Geogr 33(1):1–21
Grant R, Oteng-Ababio M (2016) The global transformation of materials and the emergence of informal "Urban Mining" in Accra. Ghana Afr Today 22:2–20. https://doi.org/10.2979/africatoday.62.4.01
Gutberlet J (2018) Waste in the city: challenges and opportunities for urban agglomerations. INTECH. https://doi.org/10.5772/intechopen.72047
Gyimah RR, Frimpong LK, Anambane G, Akolaa RA, Danso-Abbeam H, Mensah SL (2022) A comparative appraisal of environmental conditions in two urban low-income communities in Accra, Ghana. AUC Geogr 57(2):87–106. https://doi.org/10.14712/23361980.2022.8
Hartley K, Schülzchen S, Bakker CA, Kirchherr J (2023) A policy framework for the circular economy: lessons from the EU. J Clean Prod 412. [137176]. https://doi.org/10.1016/j.jclepro.2023.137176
Heath A, Swabey K (2014) Content and leximancer analysis as methodological tools to study formal school grievances: an archival study. University of Tasmania, Launceston. https://files.eric.ed.gov/fulltext/ED596767.pdf
Hill M, Hupe P (2015) Implementing public policy, 3rd edn. Sage, London
Hu J, Xiao Z, Zhou R, Deng W, Wang M, Ma S (2011) Ecological utilization of leather tannery waste with circular economy model. J Clean Prod 19(2–3):221–228. https://doi.org/10.1016/j.jclepro.2010.09.018
Hudson B, Hunter D, Peckham S (2019) Policy failure and the policy-implementation gap: can policy support programs help? Policy Design Pract. 2:1–14. https://doi.org/10.1080/25741292.2018.1540378
Hussein IAS, Mansour MSM (2018) Solid waste issue: sources, composition, disposal, recycling, and valorization. Egypt J Pet 27(4):1275–1290

International Resource Panel (2013) International resource panel work on global metal flows. http://www.resourcepanel.org/sites/default/files/documents/document/media/e-book_metals_introduction_english.pdf

Iqbal M, Syed JH, Breivik K, Chaudhry MJI, Li J, Zhang G, Malik RN (2017) E-waste driven pollution in Pakistan: the first evidence of environmental and human exposure to flame retardants (frs) in Karachi city. Environ Sci Technol 51:13895–13905. https://doi.org/10.1021/acs.est.7b03159

Kumar A, Holuszko M, Espinosa DCR (2017) E-waste: an overview on generation, collection, legislation and recycling practices. Resour, Conserv Recycl 122:32–42

Maes T, Preston-Whyte F (2022) E-waste it wisely: lessons from Africa. SN Appl Sci 4(3):72. https://doi.org/10.1007/s42452-022-04962-9

Manmohit S, Parteek ST, Siby J (2018) Health risk assessment of the workers exposed to the heavy metals in e-waste recycling sites of Chandigarh and Ludhiana, Punjab, India. Chemosphere 203:426–433

May PJ (2014) Implementation failures revisited: policy regime perspectives. Public Policy Adm 30(3–4):145–174

Mhatre P, Panchal R, Singh A, Bibyan S (2021) A systematic literature review on the circular economy initiatives in the European union. Sustain Prod Consum 26:187–202

Mugwagwa J, Edwards D, de Haan S (2015) Assessing the implementation and influence of policies that support research and innovation systems for health: the cases of Mozambique, Senegal, and Tanzania. Health Res Policy Sys 13(21). https://doi.org/10.1186/s12961-015-0010-2

Naidoo D, Nhamo L, Lottering S, Mpandeli S, Liphadzi S, Modi AT, Trois C, Mabhaudhi T (2021) Transitional pathways towards achieving a circular economy in the water, energy, and food sectors. Sustainability 13:9978. https://doi.org/10.3390/su13179978

Navarrete-Hernandez P, Navarrete-Hernandez N (2017) Unleashing waste-pickers potential: supporting recycling cooperatives in Santiago de Chile. World Dev 101:293–310. https://doi.org/10.1016/j.worlddev.2017.08.016

OECD (2020) The circular economy in cities and regions: synthesis report. https://doi.org/10.1787/10ac6ae4-en

OECD (2021) Towards a more resource-efficient and circular economy: the role of the G20. https://www.oecd.org/environment/waste/OECD-G20-Towards-a-more-Resource-Efficient-and-Circular-Economy.pdf

Ogunmakinde O, Sher W, Egbelakin T (2021) Circular economy pillars: a semi-systematic review. Clean Technol Environ Policy 23:899–914. https://doi.org/10.1007/s10098-020-02012-9

Ohajinwa CM, Van Bodegom PM, Osibanjo O et al (2019) Health risks of polybrominated diphenyl ethers (PBDEs) and metals at informal electronic waste recycling sites. Int J Environ Res Public Health 16:906

Okyere SA, Frimpong LK, Diko SK, Abunyewah M, Kita M (2021) Situating everyday urban struggles within the context of the SDGs in an informal settlement in Accra, Ghana. In: Cobbinah PB, Adanney M (eds) Sustainable urban futures in Africa. Routledge, 605 Third Avenue, New York, NY, pp 261–283

Ongondo FO, Williams ID, Cherrett TJ (2011) How are WEEE doing? a global review of the management of electrical and electronic wastes. Waste Manag 31:714–730. https://doi.org/10.1016/j.wasman.2010.10.023

Oteng-Ababio M, Amankwaa EF, Chama MA (2014) The local contours of scavenging for e-waste and higher-valued constituent parts in Accra, Ghana. Habitat Int 43:163–171

Owusu-Twum MY, Kumi-Amoah G, Heve WK, Lente I, Owusu SA, Larbi L, Amfo-Otu R (2022) Electronic waste control and management in Ghana: a critical assessment of the law, perceptions and practices. Waste Manage Res 40(12):1794–1802

Prakash S, Manhart A, Amoyaw-Osei Y, Agyekum OO (2010) Socio-economic assessment and feasibility study on sustainable E-waste management in Ghana. Commissioned by inspectorate of the ministry of housing, spatial planning and the environment of the Netherlands and the

dutch association for the disposal of metal and electrical products. https://www.ocko.dc/oekodoc/1057/2010-105-en.pdf
Preston F, Lehne J (2017) A wider circle? the circular economy in developing countries. Briefing, Royal Institute of International Affairs, London. https://www.chathamhouse.org/publication/wider-circle-circular-economy-developing-countries
Preston F, Lehne J, Wellesley L (2019) An inclusive circular economy priorities for developing countries. https://www.chathamhouse.org/sites/default/files/publications/research/2019-05-22-Circular%20Economy.pdf
Reike D, Vermeulen WJV, Witjes S (2018) The circular economy: new or refurbished as CE 3.0? Exploring controversies in the conceptualization of the circular economy through a focus on history and resource value retention options. Resources Conservation Recycl 135:246–264
Shand W (2020) The finance landscape in Ghana: mobilising investment in sustainable urban infrastructure. Paper for the coalition for urban transitions. London and Washington DC. http://newclimateeconomy.net/content/cities-working-papers
UNEP (United Nations Environment Programme) (2019) Technical guidelines on transboundary movements of electrical and electronic waste and used electrical and electronic equipment, in particular regarding the distinction between waste and non-waste under the Basel convention. http://www.basel.int/Implementation/TechnicalMatters/DevelopmentofTechnicalGuidelines/TechnicalGuidelines/tabid/8025/Default.aspx
UNEP (United Nations Environment Programme) (2013) E-waste management in Ghana (E-MAGIN Ghana): From cradle to grave. https://www.unep.org/switchafricagreen/projects/ghana/ongoing/e-waste-management-ghana-e-magin-ghana-cradle-grave
Wang S, Li J, Liu X, Zhao F, Eghbalian N (2022) Multi-level charging stations for electric vehicles by considering ancillary generating and storage units. Energy 247:123401. https://doi.org/10.1016/j.energy.2022.123401
Windapo AO, Rotimi JO (2012) Contemporary issues in building collapse and its implications for sustainable development. Buildings 2:283–299. https://doi.org/10.3390/buildings2030283

Towards just Circular Transitions in the Slums of Global South Cities

Matthew Abunyewah, Seth Asare Okyere, Michael Odei Erdiaw-Kwasie, and Festival Godwin Boateng

Abstract This concluding chapter weaves together the individual contributions in this book to develop a slum-circular economic model that aligns slum livelihoods, slum spatial appropriations, and slum housing with the core principles of the circular economy. The chapter makes the case for a just circular transition. It argues that policies that instill distributive, procedural, and recognitional justice in the slum-circular economy nexus are essential to reduce the harms and maximise the benefits of slum dwellers' everyday interactions with waste and other related materials. It teases out strategies such as social protections, participatory decision-making through coproduction, occupational health, skills training, and innovation in the policy and development planning for circular and sustainable cities. It concludes with three theoretical propositions and future research directions.

Keywords Slums · Circular economy · Synergy · Sustainability transitions

M. Abunyewah (✉)
Australian Center for Resilience Implementation and Sustainable Communities, Charles Darwin University, Casuarina, Australia
e-mail: matthew.abunyewah@cdu.edu.au

S. A. Okyere
School of Landscape Planning and Architecture, University of Arizona, Tucson, USA
e-mail: sethaokyere@arizona.edu

M. O. Erdiaw-Kwasie
Business and Accounting Discipline, Faculty of Arts & Society, Charles Darwin University, Casuarina, NT 0810, Australia
e-mail: michael.erdiaw-kwasie@cdu.edu.au

F. G. Boateng
The Earth Institute, Columbia University, New York, USA
e-mail: gb2751@columbia.edu

S. A. Okyere et al. (eds.), *Urban Slums and Circular Economy Synergies in the Global South*, Advances in 21st Century Human Settlements,
https://doi.org/10.1007/978-981-99-9025-2_11

1 Introduction

The contributions in this book add an important voice to the emerging body of work (see Brouwer et al., 2023; Rigion, 2022) contesting the single, linear, and derogative view of slums as environmentally deteriorative in the urban sustainability debate. As shown in Table 1, the evidence presented in this book argues that the principles underpinning CE are embedded in the activities, practices, and processes within slums. To frame this evidence systematically and highlight the synergies, we refer to Abunyewah et al.'s (2023) three-dimensional categorisation of the culture of slum life and processes: (i) livelihoods, (ii) space use and adaptation, and (iii) housing designs and building adjustment. Figure 1 conceptually synthesises the slum-CE synergies to establish how circularities in everyday slum life can contribute to achieving sustainable, inclusive, and circular cities. On slum livelihoods, for example, the potential to convert biodegradable waste to animal feed has been demonstrated in the Kizooka settlement of Uganda (Chap. "Circular Economy in Africa's Informal Cities: A Review of Residents' Value Retention Practices and Their Implications for Participatory Urban Planning"), exemplifying how the perception of *waste value* can support agro-based livelihoods in a region where climate change impacts such as drought make livestock rearing a fragile livelihood for the urban poor.

Housing and its services remain a fundamental aspect of slum life and culture, often a major indicator of residents' well-being. As with most slum life, housing is fluid, continuously adjusted, and modified to accommodate changing needs, rules, and aspirations (Kamalipour and Dovey, 2020). In Kibera, considered Africa's largest slum, residents use recycled plastic to replace mud as part of strengthening the structural quality and other house conditions (Chap. "Circular Economy in Africa's Informal Cities: A Review of Residents' Value Retention Practices and Their Implications for Participatory Urban Planning"). Like those in Dharavi, also labelled the world's largest slum, housing modifications or adjustments from waste are incremental processes that speak to the fluidity of informal living as in a constant state of flux. Although such processes and activities might be considered survivalist, it is important to note that a larger proportion of waste in global south cities are poorly treated or managed, generating significant costs to people and the environment. Therefore, attempts to keep products in use or design out waste are essential to reduce the environmental harms of excessive but unmanaged waste. This calls for a more expanded discussion of the slum-circular economy linkages, moving beyond a singular view of the impact of waste on slum dwellers to account for waste generation in the first place and those who contribute to its life span and value retention.

A lot of what constitutes the sociocultural life of urban slums occurs in-between spaces, such as public or semi-public spaces beyond the housing unit. Here too, everyday appropriations intersect with circular economy principles. For instance, consciously aware of their limited access to public and recreational spaces essential to the valorization of urban life, slum residents in Dandora, Kenya, have repurposed a dump site into a community garden and outdoor recreation space for children (Chap.

Table 1 Synergies between circular economy and slum

Theme	Slum-dwellers activity	Location	Related circular economy (CE) principle	Implication for circular cities	Selected book chapter
Slum livelihoods	Recycling of cardboard and aluminium waste	Belo horizonte, Brazil	Recycling	Keep products in use	Two
	The use of plastic bags as storage facilities	Rocinha, Brazil	Repurpose	Keep products in use	Two
	Basins/halved jerricans for cultivating vegetables	Kigali, Rwanda	Reuse Repurpose	Keep products in use	Four
	Turn food and other biodegradable wastes into organic manure to boost soil nutrients	Kigali, Rwanda	Reuse	Keep products in use regenerate natural systems	Four
	Buying and selling second-hand shoes, clothes, bottles, and electronic gadgets	Kigali, Rwanda	Resell	Keep products in use	Four
	Biodegradable waste is converted to animal feed	Kikooza, Uganda	Reuse	Keep products in use	Seven
	Use of agricultural residue to produce electricity	Central Uganda	Cover	Keep products in use	Eight
Slum space utilisation and adaptions	Use of rooftop for gardening	Dhaka, Bangladesh	Repurpose	Regenerate natural systems	Six
	Vertical extensions of apartments	Haifa, Israel	Refuse	Design out waste	Six

(continued)

Table 1 (continued)

Theme	Slum-dwellers activity	Location	Related circular economy (CE) principle	Implication for circular cities	Selected book chapter
	Developed parks and gardens on the dumpsites	Dandora, Kenya	Repurpose	Regenerate natural systems	Seven
Slum housing, design, and building adjustments	The use of waste materials for housing development	Dharavi, India,	Reuse	Keep products in use	Two
	Top floor of some houses without complete walls is used as water tanks, laundry, and recreational purposes	Khâm Thiên, Vietnam	Repurpose	Design out waste	Three
	Use of old corrugated iron sheets for fences to delineate plots	Maputo, Mozambique	Reuse Repurpose	Keep products in use	Three
	Use of old advertisement billboards for interior wall cladding	Khâm Thiên, Vietnam	Reuse Repurpose Refurbish	Keep products in use	Three
	Use of construction and demolition waste for pipe bedding and pavement	Dhaka, Bangladesh	Reuse Recycle Repurpose	Keep products in use design out waste	Five
	Detached houses replaced with apartment buildings/ frontage shops	Kabul, Afghanistan	Refurbish Repurpose	Design out waste	Six
	Building walls using plastic to replace houses constructed of mud	Kibera, Kenya	Reuse Repurpose	Keep products in use design out waste	Seven
	Waste as a source of fuel for cooking	Kibera, Kenya	Recover Reuse	Keep products in use	Seven

Source Authors' own elaboration

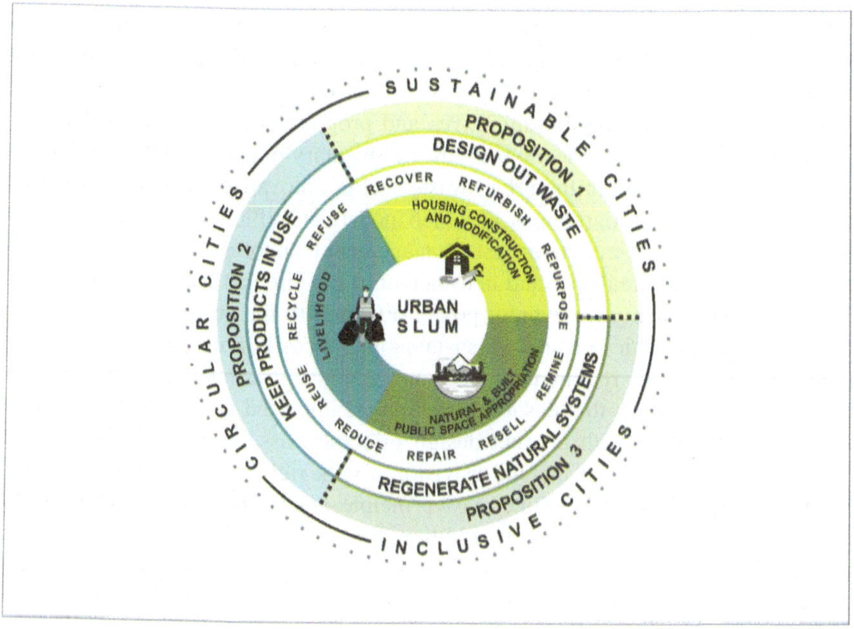

Fig. 1 Conceptual scheme of urban slum-circular economy synergy. *Source* Authors' elaboration

"Circular Economy in Africa's Informal Cities: A Review of Residents' Value Retention Practices and Their Implications for Participatory Urban Planning"). Similarly, the repurposing of rooftops for gardening initiatives has also been reported in Dhaka, Bangladesh (Chap. "Circularities in Housing Transformation Practices: A Synergetic Review"). These informal green initiatives can offer co-benefits of improving slum residents' access to green infrastructure, support the regeneration of natural systems, and, eventually, their potential contributions to sustainable development in the global south.

2 Policy Directions for just Circular and Sustainable Transitions

Based on the schematic framework (Fig. 1), this section provides three policy implications based on the recurring issues and findings across various chapters in this book. As presented earlier, the fact that urban slums have synergies with CE presents opportunities for fostering sustainable, circular, and inclusive cities in the global south. Slum formations constitute a key feature of urbanisation in many southern cities, and those living in slums are disproportionately present in the informal sector and associated livelihoods that concern material (waste) disposal and recompositions.

Here, a better alignment of slum living with CE principles in terms of policy priorities, interventions, and development programmes could provide opportunities for inclusive and equitable urban development. Therefore, policymakers should develop pragmatic and locally feasible initiatives and programmes that foster innovation. Given the reality of public health hazards, rudimentary practices, and fragile livelihoods of slum dwellers in the urban waste sector, innovative practices introduced through skill training programmes, access to modern and efficient tools, and capacity building to access cost-effective waste management technologies are necessary.

It is simplistic, unrealistic, and detrimental to the circular agenda to extol slum dwellers' circular activities without proper attention to the production of waste in the first place. To achieve a smooth sustainability transition in global south cities, significant changes are required. Global south national governments should develop comprehensive policies that ensure that products imported and produced for their citizens strictly align with product design principles associated with the circular economy. Such policies can include provisions for the right to repairability, extended producer responsibility, the polluter pays principle, and the introduction of a deposit scheme for packaging materials. In addition, the circular policy design, education, and awareness programmes need to be extended to less privileged communities to facilitate sustainability transitions of slums. Such policy imperatives cannot be taken at face value. Contributions in the volume (see Chap. "Closing the Policy-Implementation Gaps in E-waste Management: Implications for Circular Economy and Sustainability in Urban Ghana") show that in the case of Ghana, national policies and regulations on e-waste management (including reducing e-waste imports) have not translated into strategies and practical action in terms of implementation, social protections or occupational health of the slum dwellers operating in the informal waste sector. On the contrary, the experience of waste workers in Brazil (see Chap. "Linking informal settler practices and circular principles in Global South: Lessons from Asian and Latin American Cities") is instructive. Coalitions and collective organising that brings together informal waste workers, civil society organisations, and municipal governments can create platforms to recognise, support, and institute regulatory strategies that harness synergies and reduce health and environmental trade-offs.

Several chapters in this book acknowledge the critical role of slums and slum dwellers play in the sustainability of global south cities through circular practices. Yet very few cases document any involvement of slum dwellers in the design, planning, and implementation of policies and programmes that concern circular and sustainable cities' agenda. This calls for a proactive and deliberate emphasis on inclusion, equity, and justice in urban planning for circular landscapes in the urban south. Like Schröder (2020), we call for 'just circular transitions' in the sense that distributive, recognitional, and procedural dimensions are actively integrated into urban and development planning processes. For distributive justice, we emphasise a focus on access to resources and rights that optimise benefits and remove the costs associated with slum dwellers' circular practices. Therefore, there is an urgent need for regulations to ensure that the burdens of waste are not borne by those who contribute little to its generation. Furthermore, transitions from environmental or public health risk livelihoods to safer jobs should not displace income opportunities for the urban poor. On procedural justice, we assert that community leaders, groups, and slum residents should be an integral part of the urban development planning process. This should be conceived right at the start rather than added as a 'last mile' in the decision-making process. Perhaps, this is key to reducing the socio-spatial marginalisation they endure.

For us, recognitional justice for slum dwellers and their circular landscapes includes recognition of their existing and potential contributions to the achievement of circular and sustainable development objectives. Rarely do such slum dweller practices make it to the front pages of international development magazines or get enlisted in so-called best practices in the conventional global north-to-south policy transfers (Boateng et al., 2023). Changing narratives and elevating the rights, competencies, knowledge, and agency of slum dwellers call for more than civil society-led bottom-up initiatives. Hybrid interventions that manifest in top-down policy recognition and integration (from global to national levels) and scaling up of slum dweller local innovations (See Okyere et al., 2022) are fundamental to their dignity, respect, and well-being. Reasoning alongside Schröder (2020), we postulate that instilling justice in slum-CE synergies could advance progress towards the achievement of SDG 3 (Good health and well-being), 5 (clean water and sanitation), 7 (Affordable and clean energy), 8 (Decent work and economic growth), 9 (Industry, innovation, and infrastructure), 11 (sustainable cities and communities), 12 (sustainable consumption and production), 13 (climate change), 14 (life below water), and 15 (life on land).

3 Theoretical Propositions and Future Research

3.1 Slum Dwellers' Knowledge and Experience of CE Promote CE Adoption and Institutional Engagement

Waste picking and sorting have been identified as a major activity in slums supporting the livelihood of about 20 million people living in slums globally (United Nations, 2010). Plastics, cardboard, glass, metals, and electronic waste are among the materials collected and sorted out by slum dwellers. The sorted wastes are bundled up and, through the principles of circular economy (e.g., reuse, recycle, repair, repurpose, etc.), fashioned into products for housing design and adjustment, including regeneration of the natural system. The hands-on and practical experiences of slum dwellers in circular activities have deepened their knowledge and understanding in designing out waste and pollution, keeping products in use, and regenerating the natural systems (Allam and Johns, 2018; Koech and Munene, 2020). Drawing from the information deficit theory and observations in slums, we propose that an increase (decrease) in slum dwellers' knowledge and experience will positively (negatively) influence their i) commitments to promoting and adopting circular interventions and ii) level of engagement with urban planning authorities championing circularity initiatives. Therefore, future research could examine the extent to which awareness of CE principles and participation in CE planning programmes shape slum dwellers' support for circular policies and initiatives.

3.2 *Regenerative Practices in Slums Facilitate Coalition Building*

Activities that aim to maintain, restore, and improve the health of soil for plant growth, increase biodiversity, and ensure sustainable food production are gaining ground in slum and informal settlements (Abunyewah et al., 2023). For example, Kigali, Rwanda is charaterised by hills and prone to soil erosion. To reduce erosion and maintain soil nutrients in parks and gardens, residents have resorted to turning food peels, non-consumed food, and other biodegradable waste into manure (Rupf et al., 2015; Ekane et al., 2016). In some instances, community groups and non-governmental organisations have supported communities and policymakers to sustain growth and productivity in gardens and parks through regenerative initiatives. Reflecting on these, we propose slum communities with an increase (decrease) in regenerative activities will have a strong (weak) coalition with public authorities and NGOs. Consequently, future research could explore how the application of nature-based solutions in slums or informal settlements can contribute to the localisation of the circular economic agenda in inclusive and just ways.

3.3 *Local Market for Used Materials and Technology Adoption Are Key for Sustainable Circularity and Livelihood in Slums*

Several studies have shown that scavenging, sorting, and bundling waste into other products through recycling, reusing, repairing, and repurposing serve as a basis for the livelihoods of many of the slum dwellers (Sasaki and Araki, 2013; Abunyewah et al., 2023). Over the past decades, emerging studies have pointed out that circularity in slums is unsustainable due to the severe health implications associated with the rudimentary approaches adopted (Korsunova et al., 2022). Asim et al. (2012) indicated that the unsustainability of circular actions is exacerbated by insults and harassment received by circular workers from middlemen traders. The growing interest in sustaining livelihoods and circularity has shown that market availability for used products and/or items made from waste is an influential factor in achieving sustainable circularity and actions in slums (Korsunova et al., 2022; Couder and Kibuthu, 2020). In addition, the infusion of modern technology in collecting and sorting waste is important to reduce the negative repercussions of using rudimentary approaches (Katukiza et al., 2010; Suryawan and Lee, 2023). As a result of these findings from the literature, we propose that increased (decreased) demand for used materials and higher (lower) adoption of technology in waste collection and sorting will lead to sustainable (unsustainable) circularity actions and livelihoods in slums. The institutional and market dynamics that interact to shape the production, distribution, and impacts of waste as well as the treatment of slums and their informal waste workers, suggest that the political economy of waste needs to be critically examined in future research. We consider this a clarion call for urban political economists to unpack the global, national, and local systems and structures that converge to dictate the benefits and costs within the slum-circular economy nexus.

References

Abunyewah M, Ackuayi ED, Nana OA (2014) The economic dimensions of slums in the Kumasi metropolis, Ghana. Res Humanit Soc Sci 20(4):68–81

Abunyewah M, Erdiaw-Kwasie MO, Okyere SA, Boateng FG (2023) Advancing a slum–circular economy model for sustainability transition in cities of the global south. Nat Sustain 1–8

Allam Z, Jones DS (2018) Towards a circular economy: a case study of waste conversion into housing units in Cotonou, Benin. Urban Sci 2(4):118

Asim M, Batool SA, Chaudhry MN (2012) Scavengers and their role in the recycling of waste in Southwestern Lahore. Resour Conserv Recycl 58:152–162

Boateng FG, Okyere SA, Ehwi R (2023) Ordering Africa's cities: indiscipline, informal workers, evictions, the media and institutional apparatus in Ghana. Environ Plan C: Polit Space, Forthcom.

Brouwer R, Sharmin DF, Elliott S, Liu J, Khan MR (2023) Costs and benefits of improving water and sanitation in slums and non-slum neighborhoods in Dhaka, a fast-growing mega-city. Ecol Econ 207:107763

Couder L, Kibuthu S (2020) Unlocking the circular economy potential to tackle the sanitation challenge. Field Actions Sci Reports J Field Actions (Special Issue 22), 72–77.

Ekane N, Mertz CK, Slovic P, Kjellén M, Westlund H (2016) Risk and benefit judgment of excreta as fertilizer in agriculture: an exploratory investigation in Rwanda and Uganda. Hum Ecol Risk Assess Int J 22(3):639–666

Harrison ME, Sharma M (2023) Urbanisation and urban villages: an overview of slum communities in India. Urban Commons, Futur Smart Cities Sustain 109–130.

Hasan HA, Sabbar SD, Mustamin SW, Yahya M, Rahman RHA, Musa HA, Jumaing B (2023) Mapping the environmental education policies for the youth to encourage rural development and to reduce urbanisation: econometric approach. J Environ Assess Policy Manag 2350013.

Kamalipour H, Dovey K (2020) Incremental production of urban space: a typology of informal design. Habitat Int 98:102133

Katukiza AY et al (2010) Selection of sustainable sanitation technologies for urban slums—a case of Bwaise III in Kampala, Uganda. Sci Total Environ 409(1): 52–62

Koech MK, Munene KJ (2020) Circular economy in Kenya. Circ Econ: Glob Perspect 223–239.

Korsunova A, Halme M, Kourula A, Levänen J, Lima-Toivanen M (2022) Necessity-driven circular economy in low-income contexts: how informal sector practices retain value for circularity. Glob Environ Chang 76:102573

Marino G (2021) The slum economy: an alliance between the circular and the informal. Renew Matter. https://www.renewablematter.eu/articles/article/the-slum-economy-an-alliance-betwee ncircular-and-informal

McGranahan G, Satterthwaite D (2014) Urbanisation concepts and trends, vol 220. International Institute for Environment and Development.

Okyere AS, Kita M (2015) Rethinking urban informality and informal settlements growth in urban Africa: a literature discussion. J Sustain Dev Afr 17(2):101–124
Okyere SA, Diko SK, Ebashi M, Kita M (2022) Obscured innovations? inventiveness in collective infrastructure management in Accra, Ghana. Informality and the City: Theories, Actions and Interventions. Springer International Publishing, Cham, pp 555–568
Okyere SA, Frimpong LK, Diko SK, Abunyewah M, Kita M (2021) Situating everyday urban struggles within the context of the SDGs in an informal settlement in Accra, Ghana. In: Sustainable Urban Futures in Africa. pp 265–287.
Pham KT, Lin TH (2023) Effects of urbanisation on ecosystem service values: a case study of Nha Trang, Vietnam. Land Use Policy 128:106599
Quaye I, Amponsah O, Azunre GA, Takyi SA, Braimah I (2022) A review of experimental informal urbanism initiatives and their implications for sub-Saharan Africa's sustainable cities' agenda. Sustain Cities Soc 83:103938
Rigon A (2022) Diversity, justice and slum upgrading: an intersectional approach to urban development. Habitat Int 130:102691
Rupf GV, Bahri PA, de Boer K, McHenry MP (2015) Barriers and opportunities of biogas dissemination in Sub-Saharan Africa and lessons learned from Rwanda, Tanzania, China, India, and Nepal. Renew Sustain Energy Rev 52:468–476
Sakketa TG (2022) Urbanisation and rural development in developing countries: A review of pathways and impacts (No. 5/2022). Discussion Paper.
Sasaki S, Araki T (2013) Employer–employee and buyer–seller relationships among waste pickers at final disposal site in informal recycling: the case of Bantar Gebang in Indonesia. Habitat Int 40:51–57
Schröder P (2020) Promoting a just transition to an inclusive circular economy (pp. p-33). Royal Institute of International Affairs. Chatham House
Suryawan IWK, Lee CH (2023) Citizens' willingness to pay for adaptive municipal solid waste management services in Jakarta, Indonesia. Sustain Cities Soc 97:104765
UN-Habitat (2022) World cities report 2022: envisaging the future of cities. Publisher: UN-Habitat
UN-Habitat (2003) The challenge of slums: global report on human settlements. United Nations/ Earthscan, Nairobi/London
United Nations (2010) The millenium development goals report 2010. https://www.un.org/millenniumgoals/pdf/MDG%20Report%202010%20En%20r15%20-low%20res%2020100615%20-.pdf

Matthew Abunyewah is a research-focused lecturer at the Australasian Centre for Resilience Implementation for Sustainable Communities within the Faculty of Health and an Adjunct Senior Lecturer at the University of Newcastle, Australia. He received his PhD in Disaster Management and Resilience from the University of Newcastle, Australia. Matthew leverages participatory and interdisciplinary research approaches to inform government policies on disaster management and resilience, circular economy and sustainability, community development, and climate change and variability. Before joining Charles Darwin University, Dr. Abunyewah worked in the private and public sectors, contributing to policy formulation and implementation in Ghana and Australia. For example, Matthew contributed to developing the Northern Territory Circular Economy Strategy 2022–2027

Seth Asare Okyere Seth is an interdisciplinary development planner working at the intersection of international development, sustainable urbanism, community resilience, and social equity. He has extensive experience across academia, policy agencies, and non-governmental organisations conducting transformative research, leading international expert teams, coordinating international joint projects, and engaging local communities to build capacity for sustainable development. His experiences span cities and communities in Africa, Europe, Asia, and North America. Seth is in pursuit of the sustainable development of vulnerable communities within mission-focused approaches to co-producing equitable, resilient, and place-based solutions. Seth holds a Ph.D. in Urban Development Planning (Osaka University, Japan) and an MSc in Urban Planning and Policy Design (Polytechnic University of Milan, Italy). He currently serves as a Visiting Assistant Professor (University of Arizona, USA) and an adjunct Associate Professor (Osaka University, Japan).

Michael Odei Erdiaw-Kwasie Michael is an award-winning author and Lecturer in the Business and Accounting Discipline at Charles Darwin University. Dr Erdiaw-Kwasie holds a PhD in Sustainability Management from the University of Southern Queensland. His research interests focus on sustainability transitions, circular economy, climate risks & resilience, and global development. He has published over 40 scholarly outputs - books, book chapters, journal articles, and discussion papers in various fields. Dr Erdiaw-Kwasie is an Editorial Board Advisory Member for the Business Strategy and the Environment Journal. Dr Erdiaw-Kwasie has worked across policy agencies and international organisations such as Transparency International, UNDP, and Intergovernmental Forum (IGF) in various capacities.

Festival Godwin Boateng Festival is a political economist. His research interests are diverse and focus on engaged scholarship at the intersection of transport, housing, land use, pollution, climate, urban governance, and policy. Festival formerly worked as a Postdoctoral Research Scholar at the Climate School of Columbia University, New York, and as an Adjunct Professor at the Department of Urban Policy and Planning of Hunter College, New York. He currently serves on the boards of the Consortium for Sustainable Urbanization and Nature: Humanities and Social Sciences Communications and will be joining the Transport Studies Unit of the University of Oxford, UK, in the winter of 2023 as a Senior Researcher in Mobility Governance. Festival received his PhD, MSc, and BA (Hons) degrees from the Royal Melbourne Institute of Technology (RMIT), Melbourne-Australia, Roskilde University, Copenhagen-Denmark, and Kwame Nkrumah University of Science and Technology, Kumasi Ghana, respectively.

Index

A
Abandoned, 97, 106, 120
Agriculture
agricultural modernization, 144
annual crops, 143
land fragmentation, 165
perennial crops, 143
soil fertility, 135
Alterations, 96, 97, 100, 101, 103–108
Aluminum, 15, 16, 99
Appliances, 99, 104, 105
Awareness, 6, 7, 15, 47, 49–52, 55–57, 62, 65, 66, 198, 200

B
Bio-based materials, 99, 106
Building materials, 7, 35, 44, 47, 48, 74, 90, 96, 99, 102–108
Buildings
dismantling of old buildings, 99
flexible buildings, 99, 100
new buildings, 72, 74, 80, 89, 99, 105, 154
refurbishment of old buildings, 99
reuse of old buildings, 99

C
Circular city principles
recycle, 99, 105
redesign, 104, 106, 108
refurbish, 50, 56, 107, 108
remanufacture, 50
renovation, 71, 73, 90, 99, 104, 164
repair, 50, 56, 57, 104, 106, 108, 168, 184, 199
repurpose, 15, 17, 50, 60, 108, 195, 196, 199
reuse, 99, 100, 104–106, 108
Circular Economy (CE), 1, 2, 5–9, 13, 20, 27–30, 36–38, 40–42, 44, 47–52, 55–59, 62–66, 72, 73, 97–100, 103, 114, 117, 118, 131–133, 149, 150, 155, 158, 159, 164, 173, 175, 176, 185, 193–199, 201
natural capital, 100, 132, 141
resource yield optimization, 132, 141
system efficiency, 14, 48, 99
waste-energy technology, 133, 134
Cities, 1–9, 11, 12, 14–16, 20, 21, 28, 31, 32, 35, 72, 75, 95–100, 102–108, 113–118, 120–124, 149–151, 153, 155, 157, 158, 160, 162, 164, 170, 173, 174, 193–199
City authorities, 13, 14, 16, 18, 20, 21, 99, 116, 117, 121–124, 174
Conserving finite natural resources, 105

D
Degraded spaces, 97
Demolish, 71, 72, 74, 101, 105
Dismantling, 105, 186
Downcycling, 27–30, 39–41, 44
Drivers of transformations
economic benefits, 103
increasing family size, 96
socio-cultural needs, 96

S. A. Okyere et al. (eds.), *Urban Slums and Circular Economy Synergies in the Global South*, Advances in 21st Century Human Settlements,
https://doi.org/10.1007/978-981-99-9025-2

taste & preference for modern lifestyles, 97

E
Economic, 3, 12, 14, 27–30, 32, 34, 36, 41, 42, 44, 47, 48, 52, 53, 58, 59, 62, 65, 67, 72, 74, 90, 96, 99, 101, 106, 107, 114, 115, 117, 119, 121, 123, 132, 149–151, 154–158, 160, 161, 166, 167, 170, 175–177, 184, 193, 199, 200
Electronic devices, 104
Energy
electricity, 8, 131, 132, 134, 140, 141, 145
renewable resources, 133
technical residue recoverable fraction, 141
theoretical residue recoverable fraction, 134
Energy efficiency, 99
Environment, 2, 4, 7, 8, 14, 15, 21, 28, 31, 33, 36–38, 41, 42, 44, 51, 56, 65, 73, 75, 98, 99, 101, 104–107, 114, 119, 120, 122, 142, 154, 157, 160, 165, 173–176, 180, 182, 184, 185, 194

F
Fossil fuel, 99, 105, 106
Functional uses, 106

G
Global south, 1, 3, 5–9, 11, 12, 14, 15, 21, 27–29, 31, 41, 43, 44, 48, 49, 66, 131, 149, 150, 153, 173–176, 179, 187, 194, 197, 198

H
Hanoi, 6, 28, 31–34, 41, 44, 73
Housing needs, 95, 101, 102, 104
Housing production, 96, 101
Housing stock, 38, 101
Housing transformations
alterations, 96, 97, 100, 104, 108
changes in building materials and fixtures, 7, 96
extensions, 96, 97, 108
replacement, 96, 97, 108

I
Informal housing, 6, 27, 28, 33, 36, 37, 40, 44
Informal settlements, 4–6, 8, 11–21, 27, 28, 31, 32, 35, 36, 41, 42, 44, 48, 50, 59, 66, 67, 74, 96, 113–118, 120–124, 169, 200

K
Khâm Thiên, 6, 31–34, 37, 39, 40, 42, 43, 196

L
Land, 8, 13, 14, 28, 95, 98, 100, 103, 105–108, 135, 136, 143, 144, 156, 160, 162, 165, 166, 168, 199
Land use, 100, 105–108, 144, 154, 165, 166
Linear
take-use-dispose, 98
Loop, 2, 29, 97, 99, 100, 104, 106

M
Maputo, 28, 31, 34, 35, 38, 41, 44, 196
Masaka District
agricultural residues, 138
smallholder farmers, 131
surveys, 135, 137, 139
Materials
concrete, 27, 29, 35, 42, 44, 99, 102–104
durable, 90, 104, 106
long-lasting, 99, 103, 104, 107, 108
steel, 41, 42, 103
Maxaquene, 31, 33–35, 37, 38, 40, 42
Maximizing resource use, 97, 99
Minimizing waste, 97, 114, 122
Mixed uses, 106, 107
Modernist architecture, 30, 31
Modifications, 7, 77, 96, 100, 102–104, 166, 194
Multi-stakeholders, 108

P
Perception, 6, 20, 47, 49–52, 66, 119, 123, 142, 145, 194
Practices, 3–9, 11–15, 20, 21, 27–30, 36–38, 40–44, 47–49, 51–53, 57–59, 61, 62, 65–67, 71, 73, 91, 95–108, 113–118, 121, 122, 124, 125, 149, 150, 156, 160, 162, 164, 169, 175,

176, 179, 180, 183–185, 187, 194, 198–200
Principles, 1, 4–8, 11, 15–17, 28, 30, 31, 36, 37, 44, 48, 50, 65, 66, 79–82, 95, 97–99, 103–106, 108, 113–115, 117, 122, 124, 132, 133, 141–143, 149, 150, 158, 160, 169, 173–176, 180, 193–196, 198–200

R
Recover, 50, 56, 81, 124, 183, 187, 196
Recycle, 15–17, 50, 81, 85, 89, 176, 183, 184, 186, 187, 196, 199
Refurbish, 49, 50, 56, 66, 107, 108, 183, 196
Remodeling, 101, 102
Repair, 50, 56, 57, 104, 106–108, 168, 174, 184, 199
Replacement, 74, 96, 97, 101, 105–108, 158, 166
Reuse, 3, 15, 17, 20, 21, 27, 29, 30, 37, 38, 41, 44, 50, 56, 60, 64, 72–74, 76, 79–82, 99, 100, 105–108, 113, 121, 124, 176, 177, 179, 183, 184, 195, 196, 199
Rio de Janeiro, 28, 31, 33, 35, 36, 44
Rocinha, 15, 17, 31, 33–38, 40, 195

S
Solar, 30, 100, 106, 107, 120, 122, 124, 133, 158
Sustainability, 1–5, 8, 12–14, 19–21, 30, 48, 51, 63, 66, 95, 97, 104, 106–108, 114–117, 121, 122, 124, 134, 157, 158, 162, 174–176, 184, 185, 187, 194, 198

T
Tiribogo village
  Biochar, 136, 137
  class tensions, 136, 137, 142
  decentralized power system, 133
  gasification, 133, 137, 143
Transformations, 7, 8, 29, 51, 95–98, 100–108, 132, 160, 162, 164–166, 169, 174, 175, 186

U
Uganda
  energy poverty, 8
  rural electrification, 132, 133
Unused, 63, 97
Upcycling, 6, 29, 40–42, 44, 47, 51
Urban development, 5, 9, 13, 14, 21, 31, 32, 34–36, 41, 42, 44, 97, 101, 114, 115, 118, 121, 123–125, 149, 150, 155, 160, 173, 177, 198, 199
Urbanization, 4, 5, 7, 11–13, 20, 35, 71, 79, 80, 82, 89, 95–97, 116, 122, 177

V
Vegetation, 95, 103, 104, 108
Vernacular architecture, 6, 7, 27, 29–31, 41, 42, 44, 45
Vertical additions, 101, 105, 106

SPRINGER NATURE

GPSR Compliance

The European Union's (EU) General Product Safety Regulation (GPSR) is a set of rules that requires consumer products to be safe and our obligations to ensure this.

If you have any concerns about our products, you can contact us on ProductSafety@springernature.com

In case Publisher is established outside the EU, the EU authorized representative is:

Springer Nature Customer Service Center GmbH
Europaplatz 3
69115 Heidelberg, Germany

The manufacturer's authorised representative in the EU is Springer Nature Customer Service Centre GmbH, Europaplatz 3, 69115 Heidelberg, Germany. If you have any concerns regarding our products, please contact ProductSafety@springernature.com

Printed and bound by CPI Group (UK) Ltd, Croydon, CR0 4YY

30/09/2026

02225314-0005

[redacted]